Figuring Figures

We work with leading authors to develop the
strongest educational materials in statistics and social science,
bringing cutting-edge thinking and best
learning practice to a global market.

Under a range of well-known imprints, including
Prentice Hall, we craft high-quality print and
electronic publications which help readers to understand
and apply their content, whether studying or at work.

To find out more about the complete range of our
publishing, please visit us on the World Wide Web at:
www.pearsoneduc.com

Figuring Figures

AN INTRODUCTION TO DATA ANALYSIS

JON MULBERG
Department of Sociology
University of Leicester

An imprint of Pearson Education

Harlow, England · London · New York · Reading, Massachusetts · San Francisco
Toronto · Don Mills, Ontario · Sydney · Tokyo · Singapore · Hong Kong · Seoul
Taipei · Cape Town · Madrid · Mexico City · Amsterdam · Munich · Paris · Milan

Pearson Education Limited
Edinburgh Gate
Harlow
Essex CM20 2JE

and Associated Companies throughout the world

Visit us on the World Wide Web at:
www.pearsoneduc.com

First published 2002

© Jon Mulberg, 2002

The right of Jon Mulberg to be identified as author of this work has
been asserted by him in accordance with the Copyright, Designs and Patents Act 1988.

All rights reserved. No part of this publication may be reproduced, stored
in a retrieval system, or transmitted in any form or by any means, electronic,
mechanical, photocopying, recording or otherwise, without either the prior
written permission of the publisher or a licence permitting restricted copying
in the United Kingdom issued by the Copyright Licensing Agency Ltd,
90 Tottenham Court Road, London W1P 0LP.

ISBN 0130 18406 3

British Library Cataloguing-in-Publication Data
A catalogue record for this book is available from the British Library

10 9 8 7 6 5 4 3 2 1
06 05 04 03 02

Typeset in 10.5/13pt Bembo by 35
Produced by Pearson Education Asia Pte Ltd
Printed in Singapore

To the memory of my grandmother
and to Uncle John, who taught me many lessons

There is free, easy-to-use software to accompany this book. Visit
www.figuringfigures.com
for software, databases, web links and a range of useful material.

Contents

	Introduction	1
1	How to describe a group of numbers	5
2	How to see if two things are linked	24
3	How to see if two tables are linked	33
4	How to check if results are reliable	44
5	How to see if several things are linked	57
6	How to see if pairs are linked	72
7	How to judge the reliability of averages	93
8	How to present tables and graphs	106
9	Putting it all together	130
10	Summary	143
	Index	151

List of figures

Chapter 1

1	Judging levels of measurement	8
2	Typical equal opportunities question	10
3	The inter-quartile range	17
4	One number to describe many differences	17
5	Distribution of income	20

Chapter 3

6	Rank correlation	36
7	Two examples of ordinal data	38

Chapter 5

8	Fire engines don't cause the damage	61
9	Relationships between variables	61

Chapter 6

10	GNP per capita by birth rate	73
11	Positive oval scattergram	74
12	Sign of quadrants	75
13	Squeezed oval	75
14	Flat oval	76
15	Perfect correlations	76
16	Zero correlation	77
17	UK unemployment, 1990–99	80
18	Crimes committed by men aged over 16 by region in relation to high alcohol consumption, Great Britain, 1999	81

19	Crimes committed by men aged over 16 by region in relation to high alcohol consumption, Great Britain, 1999	83
20	Intercepts	84
21	Slopes	84
22	Road sign indicating steepness of slope	85
23	Positive and negative slopes	85
24	Trend of proportion of senior citizens, UK, 1951–1996	88
25	Projected trend of proportion of senior citizens, UK, 1951–1996; progection to 2056	88
26	Cinema attendances, UK, 1987–1997	89
27	Cinema attendances, three-year moving average, UK, 1987–1997	89

Chapter 7

28	Distribution of sample means	94
29	Distribution of normal curve	96
30	Distribution of normal curve	96

Chapter 8

31	Two examples of ordinal data	111
32	Cinema admissions: interest rather than simplicity	119
33	Cinema admissions: simplicity is usually best	120
34	1997 income group by 1991 income group	121
35	Energy use by selected countries	121
36	Energy use by selected countries	121
37	Income by selected organisations	122
38	High alcohol consumption by occupation and gender	123
39	High alcohol consumption by occupation and gender	123
40	High alcohol consumption by occupation and gender	124
41	Voting behaviour by religion, US presidential election, 2000	124
42	Voting behaviour by religion, US presidential election, 2000	125
43	Voting behaviour by religion, US presidential election, 2000	125
44	Trend of black smoke emissions, UK, 1970–97	126

45 Cinema admissions	126
46 Distribution of gross and net incomes, UK taxpayers, 1997–98	127
47 GNP per capita by birth rate	128

Chapter 9

48 High exam passes by RSG, 1999	134

List of tables

Chapter 1

1	GCSE entries by grade and gender	6
2	Levels of measurement and measures of central tendency	9
3	University applicants by race, 2000	10
4	Low incomes by region, England and Wales, 2000	11
5	Alcohol consumption per week in units, Great Britain, 1998	12
6	Alcohol consumption per week in units (with cumulative percentage), Great Britain, 1998	13
7	Alcohol consumption per week in units	14
8	Measures of dispersion	15
9	Gross weekly earnings by region, occupation and gender for full-time employees, UK, 1998	15
10	Distribution of incomes, UK taxpayers, 2000/01	20
11	Alcohol consumption per week by men aged 16 and over in units, Great Britain, 1998	22

Chapter 2

12	Voting behaviour by income, US presidential election, 2000	26
13	Voting behaviour by income, US presidential election, 2000	27
14	1997 income by 1991 income, UK adults	28
15	1997 income by 1991 income, UK adults (with row percentages)	29
16	Attitudes to petrol tax	30
17	1997 income by 1991 income	31

xii LIST OF TABLES

18 Voting behaviour by income category, US presidential election, 2000 31

Chapter 3

19 *The Times*, UK Universities Table 2000 35
20 University rankings by entry requirements 39
21 NHS hospital trusts by mortality rates and doctor/patient ratio, 2001 40
22 University performance by entry requirements, UK, 2000 42
23 Difference in rankings 43

Chapter 4

24 1997 income by 1991 income, UK adults 45
25 Model of no association 46
26 1997 income by 1991 income, UK adults: expected frequencies 47
27 Voting behaviour by party support 48
28 University rankings by entry requirements 50
29 Voting behaviour by income, US presidential election, 2000 52
30 Voting behaviour by income, US presidential election, 2000: model of no association 52
31 Voting behaviour by income, US presidential election, 2000: expected frequencies 52
32 Voting behaviour by income, US presidential election, 2000: residuals 53
33 Voting behaviour by income, US presidential election, 2000: chi-square 53
34 One degree of freedom, US presidential election, 2000 53
35 Minimum values of chi-square for reliability 54
36 Minimum values of Spearman's rho (r_s) for reliability 55

Chapter 5

37 1997 income by 1991 income, UK adults 59
38 1997 income by 1991 income, UK adults 59

39	University applicants by class, UK, 2000	63
40	University applicants by class and gender, UK, 2000	63
41	University applications by class and gender, UK, 2000	64
42	University applications by class, gender and rejection rate, UK, 2000	64
43	University applications by class, gender and rejection rate, UK, 2000	64
44	University applications by class, gender, rejection rate and race, UK, 2000	66
45	University applications by class, gender, rejection rate and race, UK, 2000	67
46	University applications by class and gender, standardised for rejection rate and race, UK, 2000	69

Chapter 6

47	GNP per capita by country and birth rate, 1996	77
48	Minimum values of Pearson's r for reliability	91
49	One-tailed, directional test: minimum values of Spearman's rho for reliability	92

Chapter 7

50	Alcohol consumption by adult men in the UK	97

Chapter 8

51	Low incomes by region, England and Wales, 2000	108
52	Council grant and exam performance, selected councils in England, 1997–99	109
53	1997 income by 1991 income, UK adults	109
54	British university applicants by occupation, gender and race, UK, 2000	110
55	Taxation opinion by occupation, UK adults, 2001	111
56	Rankings and entrance requirements by university, UK, 2001	113
57	Notifiable offences recorded by the police in England and Wales, by offence, April 1999 to March 2000	115

58 Direction of two-way tables – Voting by income, US exit poll, 2000	116
59 Constructing an index of senior citizens	117
60 Violent crime reported by police	117
61 1997 income by 1991 income, UK adults (with row percentages)	118
62 1997 income by 1991 income, UK adults	118

Chapter 9

63 Five 'high' GCSE passes by RSG per capita, England, 1996–99	134
64 Frequencies of significant changes in LEA pass rates, England, 1998–99	135
65 Types of probability sampling	138

Chapter 10

66 Summary of statistical tests	145

List of boxes

1	North–South divide?	11
2	Case study: hospital performance tables	40
3	Rules of thumb for category tables	48
4	Class bias in universities	62
5	Pension time bomb?	86
6	Equal pay?	99
7	Support for corporal punishment?	101

Author's note

The examples used in this book are meant only as illustrations and teaching aids. They are not meant to be used in discussions about substantive social science or policy. I have frequently taken liberties with the data, and even doctored the data on occasion to help to illustrate the aspect of data analysis I am trying to get across.

I do hope that the reader will find some of the wide range of examples interesting though, and they may wish to follow them up. I've given sources and further references to help with this. But please do not use the data in this book to justify any position in a debate, important and topical though these debates are.

J.M.

Acknowledgements

This book owes much to the patience and enthusiasm of Matthew Smith, my editor at Pearson, who has been prepared to wait while I went through a very tough time in my life. It also owes a debt to Lyn Roberts of Pearson Education, who had the good sense and sharp wit to stop selling and start buying. I also owe thanks to the anonymous referees, who were very positive. I certainly should thank successive generations of students who have had this material inflicted upon them.

PUBLISHER'S ACKNOWLEDGEMENTS

We are grateful to the following for permission to reproduce copyright material:

Chapter 3, Tables 1 and 3 from *The Times Good University Guide*, © Times Newspapers Limited, 2000; Chapter 6, Figure 9 from *www.hmso.gov.uk*, Crown copyright material is reproduced under Class Licence Number CO1W0000039 with the permission of the Controller of HMSO and the Queen's Printer for Scotland.

While every effort has been made to trace the owners of copyright material, in a few cases this has proved impossible and we take this opportunity to offer our apologies to any copyright holders whose rights we have unwittingly infringed.

Introduction

There is a sort of myth in common circulation that there are only two types of people in this world, brilliant statisticians who are plugged into computers, drink complex formulae and crunch numbers for breakfast, and the rest of us. The aim of this book is to explode this myth and make the basics of data analysis accessible to everyone.

The sort of person this book is aimed at is not a statistician. They will be able to do reasonable arithmetic, find an average or do a percentage. What they need to know is how to use these to analyse data. They may feel uncomfortable or unfamiliar with numbers and believe that statistical analysis is not for the likes of them, or maybe they even suffer from a fairly acute 'figure-phobia'. They might have looked at a textbook, seen pages of algebra and decided that this business was beyond them.

If that is you, then this book is for you. You might be a social science, psychology or management student learning research methods, or maybe a professional in a managerial post who now needs to understand data, or simply the ubiquitous 'intelligent layperson' who wants to improve their understanding of current affairs or business. Whatever the reason, this book will take you from a knowledge of basic arithmetic step by step through the fundamentals of statistical analysis.

The idea that statistical analysis is somehow only for the initiated few is similar to the view of computers about 20 years ago – that it was only for those with computer science training and a few gifted amateurs. The rest of us were tolerated but expected to know our place. Just as computing has changed, so the approach to data analysis needs to change. Actually anyone can play, and if the jargon and the algebra are stripped away what is left is fairly straightforward.

I failed my first statistics exam. It was part of a BA in management and was taught by a statistician. We learned streams of equations, and the lecturer made sure that everything was defined exactly and that we knew all the jargon. To most of us the course was not only incomprehensible but also irrelevant, in that it was not related to anything we needed to know. Ten years later, I got back on the horse and decided to try to learn statistics again. And when I began teaching statistics to undergraduates I vowed never to teach in the traditional manner I was first taught. This book reflects that philosophy.

Figuring Figures grew out of the courses in data analysis I taught at undergraduate and postgraduate levels. I was extremely dissatisfied with the textbooks that were available, all of which seemed to assume a familiarity with mathematics that my students simply did not have. What my students needed was something that bridged the gap between the textbooks and school-level texts demonstrating pie charts. I therefore wrote extensive handouts for my courses. These handouts form the basis of the present book.

WHAT THIS BOOK IS MEANT TO DO

This book takes the reader from a starting point of no knowledge about statistics to the level of understanding that most textbooks assume its readers have. It covers the material that should be in a basic introductory course (most courses cover too much). While not being the last word in data analysis, this book will help you to solve most of the problems you will meet most of the time.

Figuring Figures has a different emphasis to most statistics textbooks in that the emphasis is on the *communication* of data. Knowledge is useless unless it can be communicated. But the book is not only about communicating our analysis to others. It is also about understanding the data that other people are trying to communicate to us, to be able to 'consume' data in an intelligent fashion instead of just turning the page. However, the emphasis is also on being a critical consumer of data, who can 'talk back' to the data being presented to them.

Figuring Figures attempts to de-mystify data analysis. It avoids jargon as much as possible, and it uses simple language. It also avoids most of the equations that bedevil many data analysis textbooks. In fact, the book takes the view that knowing how to calculate the statistical tests is a bit of an obsolete skill, since cheap calculators can now do most of the calculations, and most serious analysis is now done on computers anyway. The methods of calculation are therefore relegated to appendices for the most part. Instead, the book concentrates on interpreting the meaning of the results and understanding what the various measures tell us. Surprisingly, this is what most newcomers to the subject have the most trouble with. Once cracked, the calculations become routine. The problem is actually with interpretation of the answers.

The emphasis is also very much on practical analysis. The book is practical in two ways. First, many of the *caveats* and precision of formal statistical theory have been swept aside and are mentioned only in the last chapter. This is data analysis for people who work in the real world. Second, it is practical in that it is linked to practice, in that examples are drawn from a variety of live issues in contemporary society. Data analysis is not an end in itself. It is used to find out about the world around us. In addition, most people learn faster if they can see the point of what they are learning, and it is not unreasonable to want to know what these statistical techniques can actually show us. Further examples and data sets can be found on the website.

OUTLINE OF THIS BOOK

Most data is communicated in one of two ways:

1 tables and charts, or

2 a summary number.

The book therefore looks at the different types of table and chart in turn and considers how to analyse them:

- *Frequency tables* show how to group numbers together to present a clear picture of a large set of data. This is covered in Chapter 1.
- *Two-way tables* show how to see if two or more things are associated with each other. This is covered in Chapters 2, 4 and 5.
- *League tables* are used to compare data in order. This is covered in Chapter 3.
- *Scattergraphs* are used to compare the actual magnitudes of two things, and to predict. This is covered in Chapter 6.
- Chapter 8 will also show how to present your own tables and charts.
- A *summary number* is what is usually meant by a 'statistic'. Each chapter will show how to match a statistic to the relevant table and will show how these are used in 'live' examples.

This book is short. I reckon that you want to get on and analyse data rather than read lots of my text. Go figure.

Chapter 1

HOW TO DESCRIBE A GROUP OF NUMBERS

- **Title**   Measures of central tendency
- **What are they?**   They indicate the middle of a group of numbers
- **Why use them?**   They help you to describe and get a handle on a large quantity of numbers quickly and easily

The following extract recently appeared in the national press:

> Graduates have never had it so good. The average starting salary for 1999 graduates was £17,500, a 4.8 per cent increase on 1998 and four times the rate of inflation. This year's will be taking home around £18,000....
> According to the NUS, the average 1998 graduate left university with debts of £4,500. Under the current loans system, once you start earning a salary of £15,000 or above, you have to start paying back your loan. And you've only got five years to do it.
>
> *The Guardian*, 26 February 2000

There are two things that might be noted about this extract. First, simply describing data – such as what graduates earn – can be very worthwhile (in this instance, the wages are higher than some of the lecturers who are teaching these students). More to the point from the perspective of this book, the extract shows the use of *summary numbers*; it uses a single number to give an idea of what all graduates owe and earn.

You could refer to graduate debt or graduate earnings in this example as a *variable*. All that means is that it is something that changes. Each of the graduating students would be referred to as a *case*. So a variable varies from case to case. What you want to be able to do is describe the whole variable, containing all the cases that are being looked at. There are two main methods for doing this: using a table and using a summary number.

FREQUENCY TABLES

One simple method of communicating a large quantity of numbers is to group the data and report how many cases there are for each group. This is called a frequency table (see Table 1). These sorts of table are commonplace and fairly easy to grasp. They simply tell the number of instances of each group − just over 155,000 women pupils gaining top grades in 1999–2000 − and so reduce the quantity of numbers that are needed to describe the variable.

Table 1 GCSE entries by grade and gender

Grade	Female	Male	All
A*	155 290	100 229	252 168
A	354 948	254 635	613 975
B	513 011	419 877	931 926
C	671 074	631 171	1 304 697
D	482 508	525 524	1 008 673
E	307 807	379 244	685 240
F	166 382	227 547	394 698
G	74 872	105 647	175 421
U	47 142	65 013	115 120
Total	2 773 033	2 708 887	5 481 920

Source: adapted from *Independent*, 24 August 2000

MEASURES OF CENTRAL TENDENCY

Let us take this logic of reducing numbers further and get a single number to describe the entire group. If they were asked to choose one number to do this, most people would instinctively select some sort of middle number. This is why the descriptions of groups or lists of numbers are called *measures of central tendency*. They use one number to describe many numbers, and that number will be the 'centre' number of the group.

Many people already have an idea of an 'average', which is why the term is used in newspapers. However, consider why the following statements are all different:

- The average annual income in the UK is £18,112.
- Most employees in the UK earn less than £12,000 a year.
- The average person in the UK earns £17,500 a year.

The reason these statements are all different, even though all are true, is that the averages they describe are all different. The first is what is generally known as an average, the aggregate of all incomes divided by the number of cases. The second refers to a 'league table' of incomes ranked from highest to lowest, and the average refers to the middle rank. The third statement alludes to some notion of a typical person and reports what they earn.

Each of these statements measures the data in a different way. In fact, there are several levels of measurement, and variables at each level need to be described differently.

Levels of measurement

This book will consider three levels of measurement:[1]

1 **Nominal**:   categories

2 **Ordinal**:   'league tables'

3 **Ratio**:   magnitudes.

Nominal variables are the lowest level and are variables that contain only categories: male/female, north, east, south, west, and so on.

Ordinal variables can be ranked in order from highest to lowest, like a league table. Many survey questions ask you to rank, say, satisfaction or agreement on a scale, perhaps from 'strongly agree' through to 'strongly disagree'. This means that the answers can be ranked in order, but there is no sense in which 'strongly agree' is worth double 'slightly agree' or whatever. At the time of writing, Manchester United football club are the runaway league champions. While my colleague may suggest in conversation that they're twice as good as the runners-up, he doesn't mean it literally; 'twice as good' cannot be measured here.

Ratio variables are magnitudes, numbers as we usually think of them. Here one case can be twice the size of another, so I (if I'm lucky) can have double someone else's income, and someone (unfortunately) can have half my age.

Sometimes the level of measurement is not clear, especially since we frequently use numerals for categories when collecting data, and often use them for ranks (1 for first, 2 for second, and so on). The thing to ask is if it makes sense to talk about one case being double another, or if there is a highest and a lowest (see Figure 1).

Note that you can move down a level from ratio to ordinal, or from ordinal to nominal, but not vice versa. So you could treat ratio

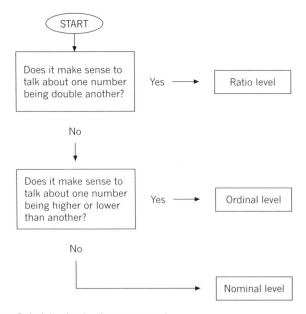

Figure 1 Judging levels of measurement

numbers as if they were only in rank order, or group an order into a couple of categories and treat it like a nominal variable, but you cannot take simple categories like region and treat them as if they were in an order. You might decide, for example, that although the data has been measured as a magnitude, the process of measurement was so dubious that it would be better to treat it as an ordinal variable.[2] By the same token, you might wish to collapse an order into two categories to present the data in a simple table. So you can drop levels of measurement, but you cannot move up.

So when we talk of 'average' income we are looking at ratio-level data: it is perfectly sensible to talk about halving your income. If you say that half the people in the UK earn above £12,000 p.a., then you are using ordinal data: you have put the incomes in order from highest to lowest. And if you talk of an 'average' person, you usually mean some sort of 'typical' wage earner, the most frequent wage earner. This is a nominal variable, since it involves only categories.

Measurement Levels and Averages

Each of the levels of measurement therefore has a different method of description; a different measure of central tendency (see Table 2).

The *mode* is the most frequent category. A nominal variable can only be described in this manner, since all you have is categories. So, for example, Figure 2 is a question from a typical equal opportunities monitoring form. This variable (ethnic origin) is a nominal variable. You cannot have double your ethnic origin; that would have no meaning. The replies cannot really be put in an order either. The replies are only categories.

Table 2 Levels of measurement and measures of central tendency

Level	Measure	What it is
Nominal	Mode	Most frequent
Ordinal	Median	Middle case
Ratio	Mean	'Average': aggregate divided by cases

Please tick the appropriate ethnic group of which you are a member:

☐ Black Carribean ☐ Indian
☐ Black African ☐ Pakistani
☐ Black other ☐ Chinese
☐ Bangladeshi ☐ Asian other
☐ Other ethnic group ☐ White

Figure 2 Typical equal opportunities question

Table 3 UK university applicants by race, 2000

Racial Origin	Applicants
Bangladeshi	3 013
Pakistani	10 223
Indian	17 024
Chinese	3 594
Asian other	4 699
Black Carribean	4 370
Black African	7 517
Black other	2 724
White	298 395
Other	7 171
Unknown	30 361
Total	389 091

Source: adapted from UCAS website

The data for university applicants is summarised in Table 3. The *modal* racial origin for our university applicants is 'white'; this is the most common reply. If you had to choose a case at random and guess what the race of the applicant was, this would be our best guess. This is why the other groups are referred to as 'minorities'.

The *median* is the middle of a league table. It is the value that has the same number of cases above and below. So the median of, for example, low income rates in different regions would be the middle region when ranked in order (see Table 4). The median rate in Table 4 would be the average of the fifth- and sixth-highest regions – the South West and the East Midlands – with a rate of 10.35 per cent of the residents earning less than £200 a week. Half of the regions in the sample have rates above (or below) this.

Table 4 Low incomes by region, England and Wales, 2000

	Gross incomes below £200/wk (%)
Wales	12.2
North East	12.0
Yorkshire & Humberside	10.6
North West	10.5
East Midlands	10.4
South West	10.3
West Midlands	9.1
East	7.9
South East	7.4
London	5.1
Total	9.2

Source: adapted from ONS website

BOX 1 NORTH–SOUTH DIVIDE?

City-based regions could be the answer

Just over a year ago the prime minister challenged the idea of a north–south divide. Some of his arguments were indisputable: the extraordinary prosperity to be found in parts of even the poorest regions; the pockets of poverty in the most prosperous regions.

Where we disagreed was with his assertion that 'the disparity within regions is at least as great as that between them' and his suggestion there should be 'a more even-handed debate'. The first ignored the depth and spread of deprivation in the north, while his even-handed debate posed the threat of even-handed help that would only widen the divide.

Now, specially commissioned research for a Guardian North debate in Salford tonight shows the economic divide between London and the once powerful northern cities has widened over the past four years and shows little immediate sign of narrowing.

Professor Brian Robson, a long-standing government adviser and head of the Centre for Urban Policy Studies at Manchester University, has compiled an array of indicators – GDP, unemployment, mortality rates, concentrations of deprivation and housing – all of which show just how far the north lags behind the south.

Leader, *The Guardian*, 1 March 2001

The *mean* is what most people would refer to as an 'average'; it is the aggregate divided by the number of cases. So the mean income for UK taxpayers in 1997–98 was £461 billion (the aggregate income) divided by 26.2 million taxpayers, giving a mean annual income of about £17,500.[3]

Describing a Frequency Table

So far, then, you've looked at the different levels of measurement and seen how to describe groups of numbers at each level. You have also seen how to generate a frequency table. What comes next is the other aspect of communication – interpreting other people's data. Consider Table 5, which shows how much alcohol a sample of adults in the UK consumed during the previous week.

The modal consumption is straightforward; it is the most frequent group in the table. So for women it is 1–7 units (which can be taken as 3.5), and for men it is 1–10 units (which can be taken as five). You could think of this as a 'typical' level of consumption.

To find the median, you need to 'count down' the table to find the middle case. To help with this, putting a running total of the percentages in the right-hand column can be a good idea. This is called a *cumulative percentage* and is in italics in Table 6. This cumulative percentage is helpful, because now all you need to do is to look for the 50 per cent mark. This is the median figure, the number that half the cases are below and half above. In this example,

Table 5 Alcohol consumption per week in units, Great Britain, 1998

Men aged 16 and over		Women aged 16 and over	
Alcohol units consumed	Frequency	Alcohol units consumed	Frequency
0	469	0	1 007
<1	514	<1	1 450
1–10	2 325	1–7	2 770
11–21	1 339	8–14	1 177
22–40	1 695	15–29	1 093
Total	6 342	Total	7 497

Source: adapted from *Living in Britain 1998*, HMSO (1999)

Table 6 Alcohol consumption per week in units (with cumulative percentage), Great Britain, 1998

Men aged 16 and over			Women aged 16 and over		
Alcohol units consumed	Frequency	Cum %	Alcohol units consumed	Frequency	Cum %
0	469	7.4%	0	1 007	13.4%
<1	514	15.5%	<1	1 450	32.8%
1–10	2 325	52.2%	1–7	2 770	69.7%
11–21	1 339	73.3%	8–14	1 177	85.4%
22–40	1 695	100.0%	15–29	1 093	100.0%
Total	6 342		Total	7 497	

Source: adapted from *Living in Britain 1998*, HMSO (1999)

the median falls in the 1–10 bracket for men and 1–7 bracket for women. Again, take the middle of this bracket as the median: five units for men and 3.5 units for women.

The mean of a frequency table is a bit trickier. What you do is assume that every case in a group has the value of the midpoint of the group. You then multiply the midpoints by the frequencies and total them up. This gives the aggregate, which is divided by the number of cases (N). So in Table 7, the midpoints are shown in column 1. You get the aggregate by multiplying these midpoints with the frequencies (column 2). This gives the results in column 3, which is shaded. The total at the foot of this column is the aggregate consumption. Divide this by the number of cases, which is the total of the frequencies in column 2. The mean consumption for women is 6.32 units (47,413/7,494), and for men 13.32 units.

Measures of dispersion

You now know how to obtain measures of central tendency that can be used to describe groups of numbers, even frequency tables compiled elsewhere. However, to complete the picture you also need to have some idea of how good a description the measure is: that is, how well the particular figure fits the numbers in the group. These measures of dispersion are summarised in Table 80.

Consider Table 9, which shows the mean earnings of full-time workers in the UK, split up into gender, occupation and region.

Table 7 Alcohol consumption per week in units, Great Britain, 1998

	Men aged 16 and over				Women aged 16 and over		
Units	[col 1] midpoint	[col 2]	[col 3] = [col 1] × [col 2]	Units	[col 1] midpoint	[col 2]	[col 3] = [col 1] × [col 2]
0	0	469	0	0	0	1 007	0
<1	0.5	514	257	<1	0.5	1 450	725
1–10	5	2 325	11 625	1–7	3.5	2 770	9 695
11–21	15	1 339	20 085	8–14	11	1 177	12 947
22–40	31	1 695	52 545	15–29	22	1 093	24 046
Total		6 342	84 512	Total		7 497	47 413

Source: adapted from *Living in Britain 1998*, HMSO (1999)

Table 8 Measures of dispersion

Level of measurement	Measure	What it is	What it shows
Nominal	None-report % of category		
Ordinal	• Range	Difference between highest and lowest case	Spread of all data
	• Inter-quartile range	Difference between medians of top and bottom half of data	Spread of middle half of data
Ratio	Variance used to get: • Standard deviation and • Coefficient of variance	Mean of squared differences Square root of variance Standard deviation / mean	Describes difference between the cases and the mean

N.B. If the coefficient of variance is high, or if the data is skewed (see below), consider using ordinal measures

Table 9 Gross weekly earnings by region, occupation and gender for full-time employees, UK, 1998

	Average gross weekly earnings (£)			
Region	Manual male employees	Manual female employees	Non-manual male employees	Non-manual female employees
London	366.70	242.10	645.90	420.10
South East	337.40	223.20	525.20	340.80
East	336.00	218.80	483.20	323.60
North West	323.70	207.30	480.10	303.70
West Midlands	327.30	203.70	471.60	303.50
Scotland	322.60	201.10	462.30	298.20
South West	309.30	202.30	461.20	304.50
East Midlands	320.50	203.80	456.00	293.00
Northern Ireland	284.40	182.40	444.00	297.40
North East	319.10	206.80	442.60	290.20
Yorkshire/Humberside	316.30	197.00	440.50	300.70
Wales	327.00	210.30	431.90	301.90
UK average	324.57	208.52	481.35	316.31

Source: adapted from ONS website

The top row shows the mean wage for all the regions: it is the mean of all the numbers below it. Note though that *none* of the regions ever matches the mean *exactly*; there is always some difference, no matter how small, between the mean and the numbers that made up the mean. For non-manual workers, both male and female, this difference between the average and each case is substantial: for male non-manual workers in London there is one-third difference between the UK average (£481) and average London earnings (£645). For London female non-manual workers the difference is also large, with a UK average income of £316 and a London average of £420.

Even if you took the median, only one case would correspond to it: all the others, by definition, would be above or below. So in a sense the idea of variation is built in to the concept of central tendency. If you are going to use one number to describe many, it follows that the vast majority of numbers are likely to be different from the single number that is used as a description. What measures of central tendency tell us is how wide the variation is between the 'central' number and all the numbers in the group.

As with measures of central tendency, dispersion measures also vary from level to level. For nominal-level data, there is no real notion of 'dispersion' other than the percentage of 'modal' cases – cases that fit the mode. This depends on the number of categories chosen, though, so it may not be too reliable.

For ordinal-level data, one obvious measure is the *range*. This is simply the difference between the highest and lowest cases. A more sophisticated measure is the *inter-quartile range*. The two quartiles are the values that are halfway between the median and the highest and lowest cases. If you like, you can think of them as being at the medians of the top and bottom halves of the group of numbers. The difference between the two quartiles is the inter-quartile range. It describes the middle half of the data for comparison with the middle case – the median (see Figure 3). In Table 9, the median weekly income across the UK for male non-manual employees is £461.75 per week (average of £462.30 and £461.20). The inter-quartile range is £481.65 – £443.40, which is £38.35. So in half the regions male white-collar employees averaged above (or below) £461, with earnings in the middle half of the regions being within £38 of this.

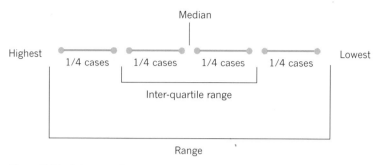

Figure 3 The inter-quartile range

Ratio-level data has a variety of possible measures. In a sense, this is a similar problem to that for the mean, except that instead of one number describing many, we're trying to get one number to describe a large number of differences between the mean and each case in our data, as shown in Figure 4. One simple way of doing this is to 'average' out these differences — to get a mean of the differences.[4] This might be called the 'mean deviation', or some such. While this is an acceptable measure, note that the problem hasn't really been solved. The only change is that there are now two means instead of one — there is the data mean and the mean of the deviations. You could ask how good a description of the deviations the mean deviation is. You could even get a new set of differences and average them, and so on *ad infinitum*.

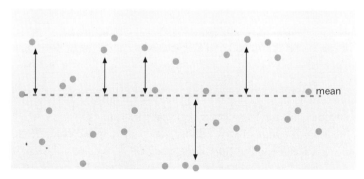

Figure 4 One number to describe many differences

Instead of the mean deviation, the usual measure of dispersion is known as the *standard deviation*. In this measure the differences are squared, and the squares are averaged out to give a measure called the *variance*. The standard deviation is the square root of the variance,[5] so for a standard deviation of 1–5:

		difference case − mean	difference squared
	1	1 − 3 = −2	4
	2	2 − 3 = −1	1
	3	3 − 3 = 0	0
	4	4 − 3 = 1	1
	5	5 − 3 = 2	4
Total	15		
	mean = 15/5 = 3	Total	10

The variance is the 'average' square of the differences, which is 10/5, or 2. The standard deviation is the square root of 2, which is 1.41. This method has the property of giving more emphasis to large deviations from the mean and is regarded as a superior description of variation.

Most scientific calculators will do standard deviations, so it is not necessary to memorise the calculation technique. The method is only included to explain where the statistic comes from.

Calculating the standard deviation of a frequency table is more complex and has been put in the appendix to this chapter. However, the result tells you the same thing, which is how varied the data is from the mean, and therefore how good a descriptor the mean is. Returning to Table 9, while the mean for male non-manual workers in the different regions is £479, the standard deviation is £56, so there is some variation between regions. The mean for male manual workers on the other hand is £324 and the standard deviation £18, a lower regional variation.

One final measure that is seldom used, although I don't really know why, has the grand title of the *coefficient of variance*, although if you just called it the variation or variation factor I reckon most people would get the idea. All it stands for is the ratio of the mean to the standard deviation, expressed as a percentage ('coefficient' is just a fancy word for a multiplication factor). So it is

100% × standard deviation / mean

This gives a 'frame' for the variation; clearly a variation factor of 5 per cent is less dispersed than one of 90 per cent.[6] In the example the variation in income between regions is 11.7 per cent for male non-manual workers and 11.4 per cent for female non-manual workers: that is, the dispersion is almost the same.

So you can now condense the description of a large set of numbers down to two – the mean and the coefficient of variance. They tell you what the centre is and how spread away from the centre the set of numbers is.

SKEWNESS

- **Title** Measure of skewness
- **What is it?** Measures discrepancy between the different measures of central tendency
- **Why use it?** It shows which *direction*, if any, ratio data varies from the mean, and the extent to which the data is not symmetrical

Another way in which the mean may not be a good description of a set of ratio data is if the variations occur in a particular *direction*: that is, most of the data is either higher or lower than the mean. Let us look again at the income distribution frequency table (Table 10). This data has a mean of around £20,800. You could present this table graphically using a histogram, such as in Figure 5. The mean and median are marked on the graph, and the mode is the highest column. It can be seen how the data is 'lopsided' – it has many more cases on the right of the chart than on the left. The mean is clearly not telling the full story regardless of the standard deviation, because that deviation is all on one side. In this case it has a skewness of 73 per cent, which is fairly high.[7] As the data becomes more skewed, the mean, median and mode will diverge.

FIGURING FIGURES

Table 10 Distribution of incomes, UK taxpayers, 2000/01

£ p.a.	thousands
4 000– 4 999	800
5 000– 7 499	3 800
7 500– 9 999	3 500
10 000–14 999	6 300
15 000–19 999	4 600
20 000–29 999	5 000
30 000–49 999	2 700
50 000–99 999	900
100 000+	300
All	27 999

Source: adapted from *Social Trends* 31 (2001)

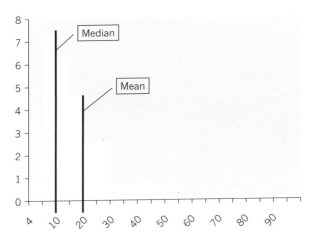

Figure 5 Distribution of income

If data is highly skewed, or if the coefficient of variance is very high, you should consider using ordinal measures of central tendency and dispersion (median, range, inter-quartile range) instead. This may give a better description. So, in the example above, saying half the cases were below £10,000 and the middle half were between around £8,000 and £18,000 may give a better idea of what is going on here.

SUMMARY

- This chapter has shown how to describe and communicate a set of data.
- Tables are the usual way of communicating data.
- Numerical descriptions of data involve using one number to describe many numbers. The simplest way to do this is to choose a 'middle' number, so they are called measures of central tendency.
- Describing the dispersion of data is as important as describing the centre of the data.
- Variables are one of three levels of measurement, and the correct measure of central tendency and dispersion should be picked to match the level of data. You can move down a level but not up.
- If the data is highly dispersed or highly skewed in one direction, consider moving down a level of measurement.

APPENDIX

Obtaining a standard deviation from a frequency table

The mechanics of getting a standard deviation from a list of numbers is outlined in many textbooks, so I won't repeat it here. Anyway, standard deviations can now be done on many pocket calculators. What I will turn to instead is the calculation of a standard deviation from a table.

As was seen earlier, tables are the main tool for communicating data. In our previous example in Table 5, you used a frequency table to calculate the mean of a sample of over 6,000. To do this, you assumed that all the cases in each category had the midpoint value. To calculate the standard deviation you use a similar technique but use the square of the midpoint instead. The mean of the data in table 11 data is 13.33.

Table 11 Alcohol consumption per week by men aged 16 and over in units, Great Britain, 1998

Alcohol units consumed	midpoint [1]	frequency [2]	freq × mid [3] = [1] × [2]	mid² [4] = [1] × [1]	freq × mid² [5] = [2] × [4]
0	0.0	469.0	0.0	0.0	0.0
<1	0.5	514.0	257.0	0.25	128.5
1–10	5.0	2 325.0	11 625.0	25.0	58 125.0
11–21	15.0	1 339.0	20 085.0	225.0	301 275.0
22–40	31.0	1 695.0	52 545.0	961.0	1 628 895.0
Total		6 342.0	84 512.0		1 988 423.5

Source: adapted from *Living in Britain 1998*, HMSO (1999)

To obtain the standard deviation, first obtain the variance by dividing the total of column 5 by the total of column 2 (N), the two shaded cells in Table 11. This gives 313.53. Then subtract the mean squared (13.33 × 13.33 = 177.69) to give 135.84. The standard deviation is the square root of this: 11.66.[8] So you can describe this data by saying that the mean consumption of alcohol is 13 units a week, with a standard deviation of 11.5 units, indicating a high variation: a factor of 88 per cent.

When using a pocket calculator to calculate standard deviation you can get some very large numbers, which may not fit on your calculator. To avoid this, you can divide the figures in either the first or second columns, or both – the values or the frequencies. So if you were looking at annual income for the UK, you could divide the incomes by 1,000 and work in 'units' of £1,000. You could also divide the frequencies by a million and work in units of a million. This should make the numbers small enough to fit on most decent calculators.

Notes

1 Some textbooks also refer to *interval* data, between ordinal and ratio. These are numbers where we cannot say one value is double another, but the gaps are constant. So 20°C is not twice 10°C (convert them to Fahrenheit), but the gap between 20°C and 10°C is the same as the gap between 40°C and 30°C. There are precious few examples of this in social science, however, so this book will not include this level of measurement.

2 I did this in my research on school league tables, which is discussed in Chapter 9. The research used a government grant as a measure of need, but the level of grant was subject to political shenanigans. It made more sense to put the data in 'league table' format.

3 Note how you can work in units of a million, to make the numbers smaller and easier. Survey data can generate some large numbers, so it is a good idea to get used to dealing with them.

4 We would have to switch all the values to positive, however, otherwise our average would always be zero: there are differences above and below the mean. Mathematicians call this the 'absolute' of a number.

5 We will ignore differences between sample and population standard deviation, because sociologists deal with large samples anyway, but if you deal with small samples you need to deduct 1 from N when calculating the variance. There is more about sample size in later chapters.

6 Note, incidentally, that because of the squaring business in standard deviations, the factor can go above 100 per cent in a set of very widely spread data.

7 The skewness is calculated by

$$3 \times (\text{mean} - \text{median}) / \text{standard deviation}$$

and as usual I convert it to a percentage by multiplying by 100. If the result is negative the data has a negative skew, and the mean is lower than the mode (the mode is the highest point on the graph). If the result is positive the mean is higher than the mode.

8 Again, for small samples the procedure is slightly different.

Chapter 2

HOW TO SEE IF TWO THINGS ARE LINKED

- **Title** Measures of association
- **What are they?** They measure the extent to which two variables change in tandem
- **Why use them?** They show if one variable affects another

In the previous chapter, you saw how to describe a single variable. You used a frequency table to group numbers together and measured both the centre and the dispersion of the data to get two numbers to describe a group of numbers.

If you only consider one variable at a time, this is really all you can do. Quite often this will be sufficient, and all you will want to do is to describe the data – how much income, the gender ratio or whatever. It is also usual to begin a report on data by briefly describing the characteristics of the data set – possibly such things as age, gender or occupation.

However, a large and very important part of social science and policy analysis is the investigation of *associations* between variables: to see what effect one thing has on another. Often you will want to examine the effect on something socially determined, such as income, crime or education; or of something 'given', such as sex, age or class. Or you may want to examine the effect of one socially determined variable on another, such as education on income, or social class on crime. The measures of association enable us to do this.

What this chapter will look at is the analysis of two variables. You will see how to find out if there is an association between two variables: that is, if one variable affects the other. You will also see how to measure how strong this association is: *how much* one variable affects the other.

You may recall that in Chapter 1 three levels of measurement were outlined: nominal, where only categories are present; ordinal, where the data is in a 'league table' format; and ratio, which are magnitudes. Just as each level has a different method of description, so each level has a different measure of association. This book will look at one measure for each level. Chapter 6 will look at ratio-level association, and Chapter 3 will look at an ordinal measure. This chapter will consider a very simple method of measuring association between two category variables. Sometimes this is called *bivariate analysis*, which simply means two variables (as in bicycle, bipolar, and so on). Later on we will look at associations between more than two variables, called *multivariate analysis*.

ASSOCIATION, CAUSALITY AND INDEPENDENCE

Before beginning this task, however, it is worth stressing that we are *not* saying that one variable *causes* another. Causality is much more than simple association. To say that one thing causes another is a much stronger claim than saying that one thing varies with another. In the latter case, you cannot be sure that something else is not affecting the association. Also, when you say A causes B, you are also saying that B would not happen if A did not. You cannot tell this from data; you do not know the result if something did *not* happen. All you know is what did actually happen. To establish a cause, I believe we need *theoretical* analysis in addition to the analysis of data. Statistics can tell us *what* happened but cannot tell us *why*.

But while you cannot say that one variable causes another, you can say that one variable is *independent* of another. Often this is common sense. If you were considering the relationship between sex and education, you would say that education varies with sex rather than the other way round (disregarding the possibility that

Table 12 Voting behaviour by income, US presidential election, 2000

	Income		
Voted for:	Below $20 000	Above $20 000	All
Gore	512	3 512	4 024
Bush	268	3 796	4 064
All	780	7 308	8 088

Source: adapted from *L.A. Times* exit poll

education can induce students to have sex changes). In this example, sex affects education, but education does not affect sex.

A good clue is the time order of the variables. The independent variable almost always occurs before the dependent. So we get our sex at birth, but our education comes later. Similarly, our parents' class status comes before our own and is therefore independent of it, although our class status might be dependent on that of our parents. Sex is independent of age – sex is at birth, age is now (incidentally, women live longer).

Given that you can usually figure out the direction of the relationship, analysing it is fairly straightforward. Again, you construct a table to do this, but since there are two variables you need to use a two-way table, such as Table 12.

Table 12 is the sort of table published regularly in newspapers. It is based on a US exit poll, and it relates the voting intentions of a sample of people to their income. The table has been simplified to two candidates and two categories of income (more about simplification later). The table has two variables, one (income) across the top and another (voting behaviour) down the side. Each variable in Table 12 has two categories, shown in bold. It has a grand total (8,088) in the bottom right, and subtotals (in italics in the table) on the right and along the bottom. The grand total, as with the frequency tables in the previous chapter, represents the number of cases and is called N, so N is 8,088 in Table 12. The subtotals are sometimes called *marginals*, since they are on the outside of the table. Each item of data – in the shaded centre of the table – is called a *cell*. There are four cells in Table 12, one for each pair of categories.

Since each variable has two categories, the table is called a '2 by 2' table, hence it has four cells.

To analyse a two-way table

1 Establish which is the independent variable.

2 Percentage the independent marginals (subtotals).

3 Compare percentages. The larger the difference, the higher the association.

The first step in analysing a two-way table is to establish which is the independent and which the dependent variable. In Table 12 the voting intention is being given for the present, whereas the income occurred previously. You should therefore take income to be the independent variable. The numbers in the cells are then turned into percentages, which standardises the figures between 1 and 100. However, these cells are not percentaged off by *N*. Rather, they are percentaged by the *independent marginals* (subtotals). In Table 12, these are the figures at the bottom of each column. This would give Table 13.

The point is that in Table 13 what you are interested in is the ratio of Gore voters to Bush voters – the dependent variable – not the ratio of low-income earners to higher-income earners. If you get this the right way round, all you need to do is compare the percentages. In Table 12, 65.6 per cent compares with 48.1 per cent – a difference of 17.5 per cent. Income makes a 17.5 per cent difference to voting behaviour in the US election.

Table 13 Voting behaviour by income, US presidential election, 2000 (cells % by independent variable subtotals)

Voted for:	Income					
	Below $20 000		Above $20 000		All	
Gore	512	65.6%	3 512	48.1%	4 024	49.8%
Bush	268	34.4%	3 796	51.9%	4 064	50.2%
All	780	100.0%	7 308	100.0%	8 088	100.0%

Source: adapted from *L.A. Times* exit poll

Table 14 1997 income by 1991 income, UK adults

1991 income group	1997 income group		
	Top 40%	Lowest 60%	All
Top 40%	13 873	6 493	20 365
Lowest 60%	6 351	24 195	30 547
All	20 224	30 688	50 912

Source: estimated from *Social Trends* 30 (2000)

Note that the sign in front of the percentage difference can be ignored, since the data is only at the nominal (category) level. The columns and rows are not in any particular order and could easily be reversed: −17.5 per cent means the same as +17.5 per cent. Since this is the case, it does not matter which row you look at in Table 13: the percentage difference will be the same. This is because both columns add up to 100 per cent, so the rows will be mirror images of each other. A simple way to remember which direction to do what is that you always compare in the opposite direction to that in which you calculate the percentages. You can tell which is the direction of calculation because the percentages in the cells add up to 100 per cent. So if you calculate down, compare across. If you calculate across, compare down.

Table 14 shows an analysis of UK income mobility. It compares the income of adults in the UK in 1997 with their income in 1991, to see if there is any mobility within groups; that is, to see if people are mobile between income groups.

Sometimes tables like Tables 12 and 14 are called *cross-tabulations*, since they tabulate two ways.[1] Table 14 cross-classifies the income groups of adults with that of seven years later. You can say that it classifies 1997 income group by 1991 income group. However, in this table the independent variable, 1991 income group, is along the side of the table. You therefore have to percentage using the row subtotals as in Table 15.

Since you percentaged across the rows (both rows equal 100 per cent), you compare down the columns. You have 68 per cent compared with 21 per cent, a difference of 47 per cent. You can say

Table 15 1997 income by 1991 income, UK adults (with row percentages)

1991 income group	1997 income group					
	Top 40%		Lowest 60%		All	
Top 40%	13 873	68%	6 493	32%	20 365	= 100%
Lowest 60%	6 351	21%	24 195	79%	30 547	= 100%
All	20 224	40%	30 688	60%	50 912	= N

Source: estimated from *Social Trends* 30 (2000)

that one's income group in 1991 makes a 47 per cent difference to the income group six years later.

Just as the different averages were given different names, so each measure of association is given a different name. Greek letters are often used. This 'percentage difference' measure is called *epsilon* (ε).

Creating two-way tables from higher-level data

While the tables in this chapter are designed for nominal (category) data, there are occasions when you may wish to examine ordinal (league or table) or ratio (magnitude) data using this method. There are several reasons why you might want to do this:

- doubts about the integrity of the data;
- to simplify the communication of analysis and results;
- to focus on one question.

So, for example, you might wish to present a simple table that can be read by people without a data analysis background (and who have not read this book), or by people in a hurry (perhaps managers reading a report). Alternatively, you may feel that while the figures are a slightly inaccurate measure, putting them in broad categories would be OK. For example, the following scale is common in attitude surveys:

'Do you agree or disagree with the statement "petrol taxes are too high"'?
☐ Strongly agree   ☐ agree   ☐ neither   ☐ disagree   ☐ strongly disagree

You may decide that while this data is actually ordinal, it would be better treated as nominal because, say, the respondents did not really have time to consider a complex question properly, so the measurement of the strength of feeling may be inaccurate. Or you may simply not wish to consider this factor and only want to look at who agrees and who does not. Either way, you would add together all 'strongly agree' and 'agree' responses together in your table, and also all 'strongly disagree' and 'disagree' responses, to produce two categories, agree or disagree. The 'neither' category would be omitted in this example.

The key thing is to decide what your new categories are to be. There are three common techniques:

1 deciding on the basis of the subject of investigation;

2 deciding on the basis of the scale;

3 using the average to split the data.

In the attitude survey example above, the scale clearly lent itself to the categories 'agree/disagree'. However, supposing that you wished to see whether being elderly affected the attitude towards petrol tax (the elderly may drive less and need public transport more) an obvious categorisation of respondents would then be over or under retirement age; this fits the subject of the investigation. By the same token, if you were interested in the association between viewpoint and income, you could calculate the mean income and divide the cases into above and below average. These techniques would give the sort of tables illustrated in Table 16. The tables are easy to follow and focus on one particular issue. The techniques also illustrate one way of simplifying large tables.

Table 16 Attitudes to petrol tax (hypothetical data)
'Should petrol tax be lowered?'

| View | Income | | | View | Age | | |
	Above average	Below average	All		Under 60	Over 60	All
Agree	70%	40%	65%	Agree	75%	50%	65%
Disagree	30%	60%	35%	Disagree	25%	50%	35%
All	100%	100%	N	All	100%	100%	N

Table 17 1997 income by 1991 income, UK adults

	1997 income				
1991 income	Lowest fifth	Next fifth	Middle fifth	Second fifth	Top fifth
Lowest fifth	5 041	2 722	1 311	706	403
Next fifth	2 342	3 768	2 138	1 324	611
Middle fifth	1 311	2 117	3 024	2 420	1 311
Second fifth	605	1 311	2 117	3 528	2 621
Top fifth	403	605	1 210	2 218	5 746

Source: estimated from *Social Trends* 30 (2000)

Essentially, you can either combine or omit categories in your variables.

Consider Table 17, which was the original data for Table 14. A useful technique in helping to analyse and communicate large tables like these is to reduce the number of categories. This can be done either by combining some of the categories, which is what was done to produce Table 14, or by omitting categories, such as any 'don't know' category, or a middle category. Originally, Table 12 looked like Table 18, with three voting categories: Bush, Gore and Nader. To aid analysis, the votes for Nader were omitted.

While these techniques do make the table simple to follow, they are also 'destroying' data, which may not always be the best thing to do, since it does involve some loss of accuracy. Which of these, simplicity or accuracy, is more important will always be a matter of judgement. Later chapters will consider methods of analysing large tables in their entirety, as well as examining associations between ordinal and ratio data.

Table 18 Voting behaviour by income category, US presidential election, 2000

	Income category		
Voted for	Below $20 000	Over $20 000	All
Gore	512	3 511	4 023
Bush	268	3 795	4 063
Nader	33	199	232
All	813	7 505	8 318

Source: adapted from *L.A.Times* exit poll

SUMMARY

- This chapter has shown how to investigate associations between two variables.
- Although you cannot say that one variable causes another, you can usually say that one variable is independent of another.
- To analyse two variables, a two-way or contingency table is used. The simplest version of this has two categories for both variables. This is a 2 by 2 table.
- A simple measure of association for category variables is the percentage difference (epsilon). To calculate this, percentage the independent subtotals and compare percentages.
- Large tables can be simplified by merging or omitting categories.
- Higher-level data, such as ordinal (league table) data, or ratio (magnitude) data, can be grouped into categories and then analysed. This makes the presentation of the analysis easier to understand, but it reduces accuracy.

Note

1 Sometimes these tables are also called *contingency tables,* since they show what *might* occur if something else happens.

Chapter 3

HOW TO SEE IF TWO TABLES ARE LINKED

- **Title** Rank correlation
- **What it does** Measures the extent to which two ordered (ranked) variables change in tandem
- **Why use it?** More powerful than category tables, but still easy to understand

Chapter 2 showed how you can find out if two category variables are associated. It showed how it is possible to simplify higher-level data and put it into a two-way table. However, this does involve a loss of accuracy, and this chapter will explain how to measure associations directly between variables that can be placed in an order. The basic idea is the same, however – you are looking to see if two variables change in tandem. Sometimes this is called *correlation*, meaning simply that the variables are related with each other (co-related).

ORDINAL DATA

While the term 'league tables' was used before to explain what ordinal variables are, actually an ordinal variable is any variable where the cases can be ranked in an order from highest to lowest (or vice versa). This commonly includes such data as:

- data from satisfaction or attitude surveys
- scaled responses
- index numbers

- ratio-level data where either measurement is a problem or the presentation of data needs to be simple
- ratio-level data that is skewed.

In the previous chapter, there was an example of a typical attitude survey. The responses ranged from 'strongly agree' to 'strongly disagree', and there were five choices. While you could say that there is an order to these replies, you could not say that 'strongly agree' is somehow worth double 'agree', or that it is five times as much as 'strongly disagree'. This would be nonsensical. So while you could rank the replies according to the level of agreement (or, for that matter, the level of disagreement), that is as far as it goes.

This example is straightforward enough, but others may be less straightforward or even ambiguous. Sometimes instead of being given choices, survey respondents are asked to provide a scaled response. This is common in satisfaction surveys. So, for example, respondents may be asked something like 'on a scale of 1 (worst) to 5 (best), how would you rate the following services?' Even though the respondents are being asked to give a number as a reply, many data analysts would suggest that this data is best treated as ordinal, since the respondents will arrive at the same numbers in a different manner, and it would be rash to assume that a score of 4 would mean the service is twice as good as a score of 2.

Index numbers can also be ambiguous. Many public services are given index numbers as indicators of performance. Table 19 is a sample of twenty universities drawn from a table of universities published each year by *The Times* newspaper. The full table has nearly 100 entries.

Usually, indices such as these are best treated as ordinal data and put in a 'league table' format. Although they have the appearance of ratio numbers, they are usually not valid at this level. In addition, it will be suggested later on that ratio data distributions that are highly skewed are also best treated as ordinal data. Even for genuine ratio-level data, the league table form can be used to simplify the presentation of your analysis, since it is an idea readily understood by most people.

Ordinal data, then, is any data in which the cases can be ranked from highest to lowest. The first step in any calculation of ordinal

Table 19 *The Times* UK Universities Table 2000 (All indices out of 100)

Institution name	Overall national position 2000	Teaching quality index	Research quality index	A-level entry points index	Employment index
University of York	5	82.4	71.5	83.2	94
University College London	6	68.8	76	84	95
University of Warwick	8	61.2	77.5	86	97
Bristol	11	58	65.5	87.2	96
Durham	11	65.2	66	82.8	91
Birmingham University	16	58.4	63.5	81.6	96
Glasgow	27	48.4	53.5	80.8	95
Reading	30	43.6	61.5	67.2	97
Queen's Belfast	36	50	47.5	75.2	97
Cranfield	39	57.2	34.5	59.2	99
Stirling	42	30	51.5	65.6	97
Aston	45	37.6	47	71.2	95
University of Aberdeen	51	30.8	55	71.6	95
Kingston University	59	43.6	12	45.2	97
Nottingham Trent	65	34.8	14	58	99
Brighton	67	36.8	19.5	39.2	95
Liverpool John Moores University	73	23.2	14.5	50.4	94
Middlesex University	78	21.2	16.5	42	93
University of Glamorgan	88	36.8	7	39.2	95
Glasgow Caledonian University	93	8.8	12	50	95

Source: The Times Universities Table

association is therefore to transform the data into ranks – first, second, third, and so on.[1] All that ordinal association does is to see if these two sets of ranks match up. If the rankings match exactly, as in Figure 6(a), this would be a perfect correlation. By the same token, if the rankings are exactly opposite, as in Figure 6(b), this would also be a perfect correlation, since it is still possible to 'read off' one result given another.

This problem of the direction of the association is a new dimension for ordinal data. Category data association had no direction, but ordinal data can run in two directions. If the rankings of two

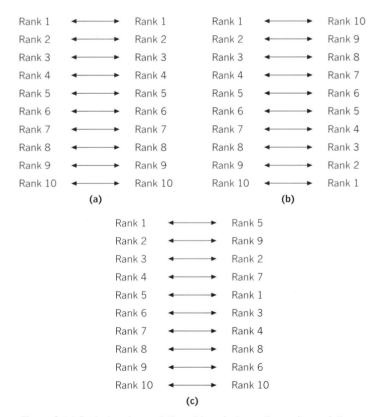

Figure 6 (a) Perfect rank correlation; (b) perfect negative rank correlation; (c) no correlation

variables run in the same direction, this is a positive correlation. If they run in opposite directions, this is a negative correlation.

If the pairs of rankings do not match up at all (as in Figure 6(c)), there is no correlation.

There are several measures of ordinal association, and many of them involve comparing ranks. The measure to be used in this book is called *rank correlation*. It was invented by a statistician called Spearman and is therefore often given his name: Spearman's rank correlation. It is sometimes given the Greek letter for r (*rho*) and denoted r_s.

The method of calculation is given in the appendix at the end of this chapter. Basically, what happens is that the rank of one variable is subtracted from the rank of the other, and arithmetic is then performed on the differences. This arithmetic yields a percentage figure, which tells how strong the association is between two variables: 100 per cent is a perfect association, and 0 per cent means no association. The sign tells the *direction* of association, not the strength: a −60 per cent correlation is just as strong an association as +60 per cent. These figures are often printed as decimals in the statistics books and in computer output: 0.6 is the same as 60 per cent.

Returning to the example of university performance indicators, one question worth considering is whether or not the performance of the university is due to its own activity, or whether it simply reflects the innate abilities of the students it admits.[2]

For the purposes of this example, the performance of the universities will be taken as being reflected in their overall position in the national league tables: the league table position is taken as a measure of their performance. The lower the number the higher the position (incidentally, Cambridge was first and York, as can be seen from the table, was fifth). The index of A-level points is an index of the entry requirements for university admission: a figure of 100 indicates the highest possible A-level examination grades. This will therefore be taken as an indicator of student ability at entry. What you need to test, therefore, is whether it is the case that the higher the A-level grade requirements, the higher the university's 'league table' position. That is, you compare the rank of the entry requirements with the ranking of the university.

The method for obtaining the rank correlation for these variables is given in the appendix to this chapter, but what you are trying to find out, broadly speaking, is the extent to which the pairs of numbers march up and down with each other. A quick eyeballing suggests that the columns both follow in order, and indeed the result for r_s in Table 20 is 92 per cent. This suggests that there is a very strong association between the ability of students before entry and the subsequent performance of the university.

It is worth bearing in mind that rank correlations can also be used for data such as attitude or customer satisfaction survey scales, or even occupational class. It may be useful, for example, to compare occupational class with attitudes that are placed on some sort of scale, as with the hypothetical question in Figure 7.

In addition, rank correlations can be used to analyse ratio-level numbers. In my own research, I looked at the data from school performance tables and investigated the association with the income each local council got from the government, which I suggested was a crude measure of need. While this data is ratio numbers, I decided to use rank correlations, for several reasons. First, the council income data was calculated for a different reason, so the actual figure may not be a completely accurate measure of need. It was also subject to variation by external pressures. In addition, my interest was in the idea of school 'league tables'. Finally, I thought it would be easier for the press to understand and report. The correlation was something like 70 per cent, suggesting that much of the exam pass data was attributable to the background of the pupils, rather than the effect of schooling.[3]

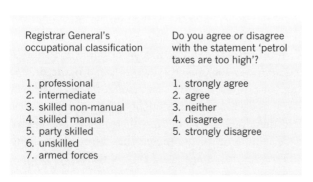

Registrar General's occupational classification	Do you agree or disagree with the statement 'petrol taxes are too high'?
1. professional	1. strongly agree
2. intermediate	2. agree
3. skilled non-manual	3. neither
4. skilled manual	4. disagree
5. party skilled	5. strongly disagree
6. unskilled	
7. armed forces	

Figure 7 Two examples of ordinal data

Table 20 University rankings by entry requirements, UK, 2000 (A-level index from 100)

Institution name	Table position	A-level points	Institution name	Table position	A-level points
University of York	5	83.2	Stirling	42	65.6
University College London	6	84.0	Aston	45	71.2
University of Warwick	8	86.0	University of Aberdeen	51	71.6
Bristol	11	87.2	Kingston University	59	45.2
Durham	11	82.8	Nottingham Trent	65	58.0
Birmingham University	16	81.6	Brighton	67	39.2
Glasgow	27	80.8	Liverpool John Moores University	73	50.4
Reading	30	67.2	Middlesex University	78	42.0
Queen's Belfast	36	75.2	University of Glamorgan	88	39.2
Cranfield	39	59.2	Glasgow Caledonian University	93	50.0

Source: *The Times* Universities Table

BOX 2 CASE STUDY: HOSPITAL PERFORMANCE TABLES

The Sunday Times recently published a study of hospital performance data in which it claimed that an association existed in UK hospitals between the number of doctors per patient and a 'mortality index'. The 'mortality index' is a complex calculation of the basic mortality rate, adjusted for a number of factors.

For the purposes of this case study, we will look at a sample of eighteen NHS trusts drawn from the national survey (see Table 21).

Because the mortality index is fairly complex, it is doubtful whether there is a simple relationship between the actual index number and the doctor/patient ratio, such that you could predict one from the other. But you could hypothesise that those hospitals with a higher ratio might also have a higher index figure. What you are trying to find out then is if the order of the two variables is the same. The result for rank correlation is that there is a 51 per cent association between the rank of the mortality index and that of the doctor/patient ratio. You can say that there is a substantial correlation between the order of the doctor/patient ratio and the order of mortality in the hospitals.

Table 21 NHS hospital trusts by mortality rates and doctor/patient ratio, 2001

Trust	Mortality index	Doctors/100 beds
Bart's and The London	70	53
Hammersmith Hospitals	88	41
North Hampshire Hospitals	88	25
Royal Liverpool and Broadgreen University Hospitals	92	35
Oxford Radcliffe Hospitals	96	70
University Hospital Birmingham	97	47
South Devon Healthcare	98	23
Epsom and St Helier	102	38
Luton and Dunstable	103	41
Queen Mary's Sidcup	103	37
Kettering General Hospital	104	22
Medway, Gillingham	106	40
Redbridge Healthcare	108	19
West Suffolk Hospitals	109	22
Countess of Chester Hospital	109	38
St Helens and Knowsley Hospitals	111	24
Burnley Health Care	112	25
Mid-Essex Hospital Services	117	31

Source: adapted from *Sunday Times* website

While the rank correlation measure does have a fairly precise meaning, the interpretation of the results remains a bit of an art. The key is to ask yourself what status you would give to a statement that a certain percentage of one variable is attributable to the relationship with something else. Consider at what level this would be noteworthy. Three-quarters of the data? Half the data? In practice, few social science associations will be much above about 70 per cent.

You have now seen how to investigate associations between two variables at the nominal and ordinal levels. However, the job is not finished. What you also need to consider is the reliability of our answers. The next chapter will look at this.

SUMMARY

- This chapter looked at association between two ordered (ordinal) variables, which can be ranked from highest to lowest.
- Measures of association compare the ranks. The better the match, the higher the association.
- If the rankings match, but in opposite directions, there is a negative correlation.
- Rank correlation gives a percentage figure that indicates the strength and direction of association. It tells how much of the dependent variable is explained by the association.
- Social science data seldom have a correlation much above 70 per cent.

APPENDIX

Calculating r_s

This chapter used an example from the UK university league tables. However, to simplify this demonstration we will look only at twelve of the entries. This gives Table 22.

Table 22 University performance by entry requirements, UK, 2000

Institution name	Table position	rank of position	A-level points (out of 100)	rank of points
Bristol	11	1.5	87.2	1
Durham	11	1.5	82.8	2
Birmingham University	16	3	81.6	3
Glasgow	27	4	80.8	4
Cranfield	39	5	59.2	6
Aston	45	6	71.2	5
Kingston University	59	7	45.2	10
Nottingham Trent	65	8	58.0	7
Brighton	67	9	39.2	12
Liverpool John Moores University	73	10	50.4	8
Middlesex University	78	11	42.0	11
Glasgow Caledonian University	93	12	50.0	9

Source: adapted from *Times* website

The first procedure is to put rankings on each index number, from highest to lowest. These are shaded in the table. If two numbers are the same, put the average of the rankings. In Table 22, Bristol University and Durham University have the equal highest national league position and are given rankings of 1.5. Glasgow Caledonian University is the lowest and is ranked 12. Similarly, Bristol has the highest entry requirement and is ranked 1.

The difference between the two rankings is then obtained for each case, and this difference is then squared and aggregated (Table 23). Note how in this table the sum of the differences beween the rankings is always equal to zero. This is a useful check that you've done the rankings correctly.

The number of pairs is N. To calculate r_s:

1 Multiply the aggregate of the differences squared (last column of Table 23) by 6.

2 Cube N ($N \times N \times N$), then subtract N from this [$N^3 - N$].

3 Divide result (1) by result (2).

4 Subtract result (3) from 1.

This gives you r_s. If like me you like to work in percentages, multiply this by 100.

Table 23 Difference in rankings

Institution name	rank of position	rank of points	difference	difference squared	
Bristol	1.5	1	0.5	0.25	
Durham	1.5	2	−0.5	0.25	
Birmingham University	3	3	0	0	
Glasgow	4	4	0	0	
Cranfield	5	6	−1	1	
Aston	6	5	1	1	
Kingston University	7	10	−3	9	
Nottingham Trent	8	7	1	1	
Brighton	9	12	−3	9	
Liverpool John Moores University	10	8	2	4	
Middlesex University	11	11	0	0	This is always zero
Glasgow Caledonian University	12	9	3	9	
Total			0	34.5	

So for the example:

1 The aggregate of differences squared is 34.5. So 6 × 34.5 is 207.

2 N is 12. $12^3 - 12 = 1,716$ [i.e. $(144 \times 12) - 12$].

3 207 divided by 1,716 is 0.12.

4 $1 - 0.12 = \mathbf{0.88}$ or 88%.

Notes

1 There may, of course, be tied ranks. To find out how to deal with these, see the appendix.

2 It could also be argued that the process is a causal loop, and that good students become attracted to the best universities. Causal loops cannot easily be investigated by statistical means, however, and this discussion will not be looked at in this book.

3 This research is discussed in greater depth in Chapter 9.

Chapter 4
HOW TO CHECK IF RESULTS ARE RELIABLE

- **Title** Significance testing
- **What it does** It indicates the reliability of the statistical measures
- **Why use it?** If the measures that are calculated are unreliable, the conclusions drawn from them may be incorrect

So far this book has considered how to describe a set of data, using one number to describe many numbers. It has also looked at how to examine an association between two variables. This chapter introduces a different question, the question of reliability. This is obviously an important question to ask. If you cannot rely on your results it would be foolish to draw conclusions from them. Part of the notion of reliability is an idea of repetition. When we say that a machine is reliable, we mean that it continues to work as normal no matter how often it is used. Similarly, if our statistical measures are reliable, a repetition of the study would yield the same result.

Another way of thinking about statistical reliability is to think whether you are able to generalise from the conclusions you reached. Consider for example Table 24, which appeared in Chapter 2 as Table 14. It can be seen from this table that there were over 50,000 respondents in the survey. The thing is that you do not really want to know about these 50,000 people. What you are really interested in is the UK as a whole. The 50,000 cases are a *sample*, drawn at random from a population, in this case the entire country. One of the most useful aspects of data analysis is precisely this ability to be

Table 24 1997 income by 1991 income, UK adults

					1997 Income Group
1991 Income Group	Top 40%		Lowest 60%		All
Top 40%	13 873	68%	6 493	32%	20 365 = 100%
Lowest 60%	6 351	21%	24 195	79%	30 547 = 100%
All	20 224	40%	30 688	60%	50 912 = N

Source: estimated from *Social Trends* 30 (2000)

able to generalise to a large group (the population) by looking at a small sample drawn from it.

Of course, in order to be able to generalise from the results of our sample, these results have to be reliable. That is, can we say that if the above study were to be repeated, and another sample of 50,000 people were to be drawn from the population, the same result would be obtained?

Before you look at this question of reliability though, there are a few points to clear up in order to avoid possible confusion. First, statisticians use the phrase *statistical significance* to describe this question of reliability. I hate this phrase and try not to use it if at all possible. It implies that the results are somehow important or noteworthy, or that the correlations are large, whereas none of these is the issue. The question of reliability is entirely separate from the question of importance or the level of association. Even if there is no association in a sample, you still need to know if the results are reliable.

Another reason for confusing reliability with association is because by doing arithmetic on the measure of reliability, a better measure of association can be obtained. In fact, the next chapter looks at this. Nonetheless, the two results measure different things, and it is important to keep the two ideas of association and reliability separate.

Reliability, then, is concerned with the relationship between a sample and the larger population from which the sample is drawn. It tells us whether the same results are likely if another randomly chosen group is sampled, since if a different result is likely to be obtained the conclusions cannot be used. You can never be 100 per cent certain of this reliability though. However many samples are drawn, the next one may yield different results. The only procedure is therefore to pick an acceptable level of reliability and test for this. By convention, social scientists use 99 or 95 per cent reliability, whereas medical or engineering levels of reliability would need to be much higher. Adopting this procedure does mean that sometimes you will disregard perfectly good data, but that is just the way things go.

MEASURING RELIABILITY: CATEGORY DATA

In general, the larger the sample or the stronger the result, the more certain you can be of the reliability. If you have a sample of 50,000 or even 10,000, the results will almost certainly be reliable, or if you have a smaller sample (say 100) but find a very high association (for example 94 per cent), then the results are also likely to be reliable.

Measures of reliability therefore consist of two main elements:

1 sample size

2 level of association.

Statisticians are strange people. They do everything backwards. These are people who always reverse into the parking space. When considering the reliability of an association, statisticians begin by considering what the data would look like if there is no association. This is called a *model of no association*.[1] Consider Table 24 again. This looks at the association between an independent variable (past income) and a dependent variable (present income). The table shows the percentages by the independent marginals (subtotals). If there is no association, however, the table percentages will look like those in Table 25. Note that the percentages in bold on the bottom row of the table – the *dependent marginals* – remain unchanged. If there

Table 25 Model of no association

1991 income group	1997 income group		All
	Top 40%	Lowest 60%	
Top 40%	40%	60%	20 365 = 100%
Lowest 60%	40%	60%	30 547 = 100%
All	40%	60%	50 912 = N

Source: estimated from *Social Trends* 30 (2000)

Table 26 1997 income by 1991 income, UK adults: expected frequencies

1991 income group	1997 income group					
	Top 40%		Lowest 60%		All	
Top 40%	8 090	40%	12 276	60%	20 365	= 100%
Lowest 60%	12 134	40%	18 412	60%	30 547	= 100%
All	20 224	40%	30 688	60%	50 912	= N

Source: estimated from Social Trends 30 (2000)

is no association, all the categories will have the same percentage as the overall variable. Epsilon would be 0. That is, the independent variable categories will make *no difference* to the dependent variable.

The complete table will look like Table 26, in which the cells are called the *expected frequencies*. These are what you would expect if there is no association. The differences between these expected frequencies and what is actually obtained (the observed frequencies) are called the *residuals*. The reliability is determined by the sample size, the size of the residuals and (loosely speaking) the number of residuals in the table.

There are two steps to the determination of the reliability of a two-way table. The first of these is the calculation of a statistic called *chi-square* (chi χ is, strangely enough, another Greek letter). The next step is to use chi-square to measure the reliability of our result. Both of these steps are covered in the appendix to this chapter. What emerges enables you to judge whether or not the association that has been discovered is reliable at the chosen level, regardless of how strong that association is.

Since by convention 99 or 95 per cent reliability is used, let us call this the higher and lower levels, respectively. You get a 'yes' or 'no' answer to the question 'is the data reliable at the higher level?' and a 'yes' or 'no' answer to the question 'is the data reliable at the lower level?' If the answer to either question is 'yes', then you can report the results and conclusions, including the level of reliability. If the answer to both questions is 'no', you can draw no conclusions from the results, which are not to be relied on.

In Table 24, chi-square is 11,430, which tells you that the data is 99 per cent reliable. So you can report that the association between the variables is 36 per cent, and that you are 99 per cent certain of

Table 27 Voting behaviour by party support, UK adults 2000, in response to the question 'Would you change your vote if you disagreed with the party you support over Europe?'

Voting behaviour	Political party		
	Con	Lab	Total
Change	53.9%	47.2%	50.0%
Not change	46.1%	52.8%	50.0%
Total	234	327	561

Source: adapted from MORI poll, June 2000

this. That is, you can be confident the results will hold true for the entire population, not just the sample.

However, consider Table 27 which is adapted from a survey on voting intentions. In this table, the reliability is less than 95 per cent.[2] This means that you can say nothing much about the relationship between the variables, regardless of what association there may be in our sample, since if you obtained another sample the results may well be different.

Again, though, statisticians do this backwards. They look at the level of unreliability rather than the level of reliability. So they say that the data is significant at the 1 per cent or 5 per cent level, which simply means that it is 99 per cent or 95 per cent reliable, respectively. Sometimes they put this result as a decimal and call it p (p for probability). So $p = 0.05$ means that the data is 95 per cent reliable.

There is one problem with the chi-square test. If the expected frequencies in any of the cells fall below 5, the test cannot be used. This is why you need to know what these frequencies are all about.

BOX 3 RULES OF THUMB FOR CATEGORY TABLES

5 If some of the expected frequencies are below 5, the results are unreliable.

400 If the sample is above 400, the results have a good chance of being reliable.

1000 If the sample is over 1,000, the results will probably be reliable.

Box 3 shows three useful rules of thumb for estimating the reliability of two-way tables. If the expected frequencies are below 5 (or if any are 0), then the results will probably be unreliable, and the chi-square test cannot be used. If the sample size is less than 400, there is a good chance that the results will be unreliable. If the sample size is above 1,000, the chances are quite good that the results will be reliable and the conclusions generalizable to the population.

MEASURING RELIABILITY: ORDINAL DATA

The mechanics of this are simpler than chi-square. Using Spearman's rank correlation, you can simply look up the result in a table. An example is given in the appendix. Again, this tells us if the correlation is reliable. So if the association is significant at the 5 per cent level, you can be 95 per cent confident that it holds for the entire population.

Table 28 is a repeat of Table 20 from Chapter 3. You will recall that it correlates performance with entrance requirements. The rank correlation for this data is 92 per cent, and there are twenty universities in the sample, so N is 20. A quick glance at the table in the appendix will show that the minimum level of r_s for reliability at the 1 per cent level (99 per cent reliability) is 59.1 per cent. The 92 per cent result is clearly above this, so the result can be taken as reliable.

There is one final point to make about reliability. A quick glance down Tables 35 and 36 in the appendix to this chapter will show that the number of cases required for reliability with ordinal data is much smaller than for nominal data, even single figures compared with several hundred for a two-way table. It would seem that results from ordinal data are much more reliable than nominal data.

The reason for this is straightforward. Ordinal data is in an order and so is much less likely to have been affected by the chance factors in random sampling. For example, your chances of calling a coin toss correctly are one in two (50 per cent). Of ten tosses, you might guess that five will be heads, which is not much of a trick. But guessing in advance the *order* of heads and tails would be quite a feat.

Table 28 University rankings by entry requirements, UK 2000

A-level index from 100

Institution Name	Table position	A level points	Institution Name	Table position	A level points
University of York	5	83.2	Stirling	42	65.6
University College London	6	84	Aston	45	71.2
University of Warwick	8	86	University of Aberdeen	51	71.6
Bristol	11	87.2	Kingston University	59	45.2
Durham	11	82.8	Nottingham Trent	65	58
Birmingham University	16	81.6	Brighton	67	39.2
Glasgow	27	80.8	Liverpool John Moores University	73	50.4
Reading	30	67.2	Middlesex University	78	42
Queen's Belfast	36	75.2	University of Glamorgan	88	39.2
Cranfield	39	59.2	Glasgow Caledonian University	93	50

Source: adapted from *Times* website

By the same token, obtaining a sample of about ten manual workers and ten non-manual workers and splitting each group by their father's occupation would not tell us much about class mobility, since the results are unlikely to be reliable. But putting the incomes of the twenty cases into a league table and seeing if they matched up with their parents' income, also in a league table, might tell us a great deal more. A high correlation in the latter instance would be much more reliable.

This is a good reason for keeping to a high level of measurement. The higher the level, the more reliable your results are likely to be for a given sample size, or the smaller your sample needs to be to achieve a given level of reliability.

Later on, you will be extending the measures of reliability to ratio data and applying it to the descriptions of single variables used in Chapter 1. However, the next chapter will use chi-square as the basis for a better measure of nominal association and look at ways of examining associations between more than two variables.

SUMMARY

- This chapter has considered a new question – the question of the reliability of results.
- Reliability is concerned with the question of whether the results from a sample can be generalised to the rest of a population. This enables us to find out about a large group by looking at a smaller number of cases.
- You can never be 100 per cent certain that if you obtained another sample the results would not be different, so choose an acceptable level – usually 99 per cent or 95 per cent certainty – and test for that.
- If the results do not match this level, you cannot rely on them, even if you find high associations.
- The concept of reliability is different to the ideas of importance of the results and levels of association.

APPENDIX

Calculating chi-square

Consider Table 29, which is Table 13 from Chapter 2. The first step is to obtain a model of no association. This replaces the cells with the same percentages as the independent marginals (Table 30). This model is used to obtain the *expected frequencies*: the frequencies you would expect if there were no association (Table 31).

The residuals are the differences between the actual data (in Table 29) and the expected frequencies (in Table 31). These are shown in

Table 29 Voting behaviour by income, US presidential election, 2000

Voted for:	Income				All	
	Below $20 000		Above $20 000			
Gore	512	65.6%	3 512	48.1%	4 024	49.8%
Bush	268	34.4%	3 796	51.9%	4 064	50.2%
All	780	100.0%	7 308	100.0%	8 088	100.0%

Source: adapted from *L.A. Times* exit poll

Table 30 Voting behaviour by income, US presidential election, 2000: model of no association

Voted for:	Income			All
	Below $20 000	Above $20 000		
Gore	49.8%	49.8%	49.8%	4 024
Bush	50.2%	50.2%	50.2%	4 064
All	780	7 308		8 088

Source: adapted from *L.A. Times* exit poll

Table 31 Voting behaviour by income, US presidential election, 2000: expected frequencies

Voted for:	Income				All	
	Below $20 000		Above $20 000			
Gore	388	49.8%	3 636	49.8%	4 024	49.8%
Bush	392	50.2%	3 672	50.2%	4 064	50.2%
All	780	100.0%	7 308	100.0%	8 088	100.0%

Source: adapted from *L.A. Times* exit poll

Table 32. The results have been rounded off for simplicity. For each of the residuals (which in a 2 by 2 table will be mirror images of each other), square the number and divide by the expected frequencies (in the example, the square of 124 is 15,376). The sum of these four calculations is chi-square (χ^2). In the example, it is 84.5 (Table 33).

To evaluate the significance of the chi-square figure, you need one other idea, the idea of 'degrees of freedom'. Broadly speaking, this is the number of cells you would need to know to complete the table given the marginal subtotals (Table 34). Once you know one cell in a 2 by 2 table, you can find the rest if you know the marginals; you can simply subtract to fill in the blanks. To get the

Table 32 Voting behaviour by income, US presidential election, 2000: residuals

	Income	
Voted for:	Below $20 000	Above $20 000
Gore	124	−124
Bush	−124	124

Source: adapted from L.A. Times exit poll

Table 33 Voting behaviour by income, US presidential election, 2000: chi-square

	Income		
Voted for:	Below $20 000	Above $20 000	Total
Gore	15 376 / 388 = 39.6	15 376 / 3 636 = 4.2	
Bush	15 376 / 392 = 39.2	15 376 / 3 672 = 4.2	
Total	76.1	8.4	χ^2 = 84.5

Source: adapted from L.A. Times exit poll

Table 34 One degree of freedom, US presidential election, 2000

	Income		
Voted for:	Below $20 000	Above $20 000	All
Gore	512		4 024
Bush			4 064
All	780	7 308	8 088

Source: adapted from L.A. Times exit poll

Table 35 Minimum values of chi-square (χ^2) for reliability

Degrees of freedom	Level of reliability	
	5% level (95% certainty)	1% level (99% certainty)
1	3.84	6.64
2	5.99	9.21
3	7.82	11.34
4	9.49	13.28
5	11.07	15.09
6	12.59	16.81
7	14.07	18.84
8	15.51	20.09
9	16.62	21.67
10	18.31	23.21
11	19.68	24.73
12	21.03	26.22
13	22.36	27.69
14	23.68	29.14
15	25.00	30.58
16	26.30	32.00
17	27.59	33.41
18	28.87	34.81
19	30.14	36.19
20	31.41	37.57

Source: adapted from R.D. Nelson *The Penguin Book of Mathematical and Statistical Tables* (Harmondsworth: Penguin, 1980)

degrees of freedom, subtract 1 from the number of rows and 1 from the number of columns and multiply these together.

(rows − 1) × (columns − 1)

A 2 by 2 table will have one degree of freedom. A 3 by 3 table will have four degrees of freedom.

You can now look up the level of significance in Table 35. Since there is one degree of freedom in the example, use the top line. If the chi-square figure is above the number in the top line (3.84 or 6.63 for 5 and 1 per cent, respectively), the results are reliable at that level. In the example, the data is therefore reliable at the 1 per cent level.

Calculating and using chi-square

1 Obtain expected frequencies (using marginal percentages). If any of these are 0 or less than 5, the chi-square test is inoperable.[3]

2 Obtain residuals by subtracting expected frequencies from what is actually obtained.

3 Square the residuals and divide by the expected frequencies. The aggregate of these is chi-square (χ^2).

4 Get the degrees of freedom by (rows − 1) × (columns − 1).

5 Look up the degrees of freedom in the left-hand column in Table 35. Then look across that row. The figure gives the minimum value of chi-square required for the respective levels of significance.

Reliability of rank correlation

To obtain the reliability of a rank correlation (r_s), simply look up the sample size N in the left-hand column of Table 36. Then read across to see the minimum values of r_s for the result to be reliable. If r_s is above that minimum figure, the result is reliable and can be generalised to the rest of the population.

Table 36 Minimum values of Spearman's rho (r_s) for reliability

	Level of reliability	
N	5% level (95% certainty)	1% level (99% certainty)
5	100%	
6	88.6%	100%
7	78.6%	92.9%
8	73.8%	88.1%
9	68.3%	83.3%
10	64.8%	79.4%
12	59.1%	77.7%
14	54.4%	71.5%
16	50.6%	66.5%
18	47.5%	62.5%
20	45.0%	59.1%
22	42.8%	56.2%
24	40.9%	53.7%
26	39.2%	51.5%
28	37.7%	49.6%
30	36.4%	47.8%

Source: adapted from Runyon and Haber, *Fundamentals of Behavioural Statistics* (Reading, Mass.: Addison-Wesley, 1967)

So to use the example in Chapter 3, there are twenty pairs of index numbers from the university performance tables, which gives a rank correlation of 92 per cent. The correlation required is 59.1 per cent for the 1 per cent level of significance (99 per cent reliability) and 45 per cent for the 5 per cent level (95 per cent reliability). Since the correlation is higher than this, the result is significant at the 1 per cent level – it is 99 per cent reliable.

Notes

1 However, there is a good reason for this reverse logic; see page 78.
2 Chi-square is only 2.48, and according to Table 35 (page 54) we need a value of 3.84 for 95 per cent reliability. See the appendix for details of calculating chi-square.
3 There is a neater way to calculate this:

(row total × column total) / overall total

Chapter 5

HOW TO SEE IF SEVERAL THINGS ARE LINKED

- **Title**   Cramer's *V*

- **What it does**   It gives a superior measure of association for category variables

- **Why use it?**   1   It can be used on large tables

    2   It doesn't require the independent variable to be worked out

    3   It is easier to interpret than simple percentage difference

This chapter will discuss two things. First, it will look at a more sophisticated measure for association between category variables than the simple percentage difference used in Chapter 3. Then this new measure will be applied to the analysis of tables with more than two variables.

Logically, discussion of the new measure belongs in the previous chapter, since the new measure is based on chi-square, the discussion of which occupied much of the last chapter. However, as I have said *ad nauseam*, a very common error for students new to statistics is to confuse the question of reliability with that of association. Putting a section on association in the same chapter as that of reliability is likely to increase that confusion. Indeed, so keen am I to avoid sowing the seeds of confusion that this chapter will not show how the new measure of association is calculated, simple though it is. This has been banished to an appendix.

In order to reinforce the difference between reliability and association, let us briefly recap the stages of data analysis covered so far:

1 Describe the data (using measures of central tendency and dispersion).

2 Analyse associations between variables if required.

3 Reliability must be checked – are the results generalisable to the population as a whole?

The question of reliability was covered in the previous chapter, and association was covered in the chapters before that. However, there are some weaknesses with the use of percentage difference (epsilon) for the analysis of category variables. While it is easy to calculate and is intuitive, it has two major drawbacks. First, it is only really useful on small tables with two categories. If there are three or more categories the procedure is no longer clear-cut. Second, it requires the analyst to ascertain which way the dependency goes – which is the dependent and which the independent variable.[1] That is to say, it is not a *symmetrical* measure. Finally, and most importantly, while epsilon is easy to calculate, it can be difficult to interpret. There is no exact meaning to the results. And since it is this question of interpretation that most people new to statistics have problems with, a more advanced measure that is easier to interpret is good news.

CRAMER'S V

The new nominal measure of association was developed by a statistician called Cramer and is called V (a welcome change from Greek letters). It is calculated by doing arithmetic on chi-square[2] but is a measure of association, not a measure of reliability. It can be used on any size of two-way table, not only 2 by 2 tables, and is a symmetrical measure, which does not require the user to work out which variable is regarded as dependent or independent – no calculating of percentages is required.

Table 37 1997 income by 1991 income, UK adults

	1997 income group		
1991 income group	Top 40%	Lowest 60%	All
Top 40%	13 873	6 493	20 365
Lowest 60%	6 351	24 195	30 547
All	20 224	30 688	50 912

Source: estimated from Social Trends 30 (2000)

Table 38 1997 income by 1991 income, UK adults

	1997 income group				
1991 income group	Top 20%	Second 20%	Middle 20%	Second lowest 20%	Lowest 20%
Top 20%	5 041	2 722	1 311	706	403
Second 20%	2 342	3 768	2 138	1 324	611
Middle 20%	1 311	2 117	3 024	2 420	1 311
Second lowest 20%	605	1 311	2 117	3 528	2 621
Lowest 20%	403	605	1 210	2 218	5 746

Source: adapted from Social Trends 30 (2000)

In Table 37, again from Chapter 2, Cramer's V is 47 per cent. That is, 47 per cent of the data for the 1997 income group can be ascribed to the association with 1991 income. Table 38 is the full table of income mobility, with incomes divided into fifths. While epsilon might be hard to calculate and interpret, V is straightforward: 67 per cent. There is a 67 per cent association between father's occupation and that of their children (chi-square is 22,664, which is 99 per cent reliable, so you can be 99 per cent certain that these results can be generalised).

The disadvantages of V are that it is harder to explain to the lay reader, since it assumes a knowledge of statistics, and is not intuitive in the way that epsilon is. It can also take longer to calculate, and in many cases it will not yield vastly different results. However, for those projects where accuracy is more important than simplicity, Cramer's V is advantageous.

SUMMARY: MEASURES OF CATEGORY ASSOCIATION

- This section introduced a new measure of association for category variables: Cramer's V.
- While epsilon is simple and intuitive, it has no straightforward interpretation. Cramer's V is based on chi-square but is a measure of association.
- It has a standard interpretation, is symmetrical and can be used on any two-way table.
- However, it cannot be as easily calculated, and it is harder to communicate.

MULTIVARIATE ANALYSIS

We can now turn to the main business of this chapter: the analysis of several variables simultaneously. This is known, rather unimaginatively, as *multivariate analysis*. This book covers multivariate analysis only at the category (nominal) level.

Let us return briefly to the discussion of correlation versus causality. You saw before how the fact of correlation does not mean that one thing *causes* another, only that they are related. There is an old joke that there is a high correlation between the damage done by a fire and the number of fire engines in attendance. This does not mean that the fire engines caused the fire. There is another variable – a prior variable – which causes both the damage and the attendance of fire engines. This is the size of the fire (Figure 8). However, there is nothing in the correlation between the damage and number of engines that would give us a clue as to where to look for another variable that contributes to the model. This requires *theoretical* knowledge, in this instance created by practical experience and common sense. In my view, data analysis in social science can never be cut off from theoretical analysis. I believe we

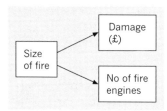

Figure 8 Fire engines don't cause the damage

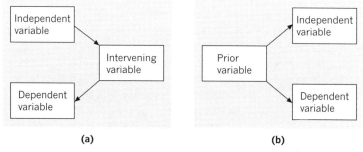

Figure 9 The relationship of (a) an intervening variable and (b) a prior variable to dependent and independent variables

need theory to know where to look for data and how to make sense of it.

The extra variables can either be in the middle of the two original variables or in front of them. Variables in the middle are called *intervening* variables; those in front are called prior variables (Figure 9). Prior variables can affect either or both of the subsequent variables. Sometimes this process of analysing additional variables is called *controlling* for the variables, and the additional variables are referred to as *control* variables.

As an example, let us look at a recent controversy over higher education.

> **BOX 4 CLASS BIAS IN UNIVERSITIES**
>
> North rejects Prescott's attacks on 'elitism'
>
> THE row over 'elitism' spread to the North yesterday when Durham and Newcastle Universities rejected criticism from John Prescott, the Deputy Prime Minister, that they were not doing enough to attract students from state schools.
>
> Mr Prescott said that the Government had no intention of backing down in its class-war assault on the country's leading universities. In a speech in Gateshead, Mr Prescott said: 'We believe many universities North and South could do more to attract people of ability regardless of their backgrounds.'
>
> Sir Kenneth Calman, Durham University's Vice-Chancellor, responded to Mr Prescott, stressing the efforts Durham had made to attract students from across all backgrounds. 'It is clear from some of the things being said by ministers that they have not been well briefed about the efforts that universities such as Durham have been making for a number of years. Durham is a world-class university. Admission is competitive and is based on academic performance and the ability to achieve. We are encouraging people from everywhere to take up these opportunities.'
>
> Prof. James Wright, Vice-Chancellor of Newcastle University, said that Government ministers were 'not really aware of the facts'. He said that Newcastle ran a partnership programme with state schools and colleges across the North East, with academics speaking to pupils about university and helping the brightest to win a place.
>
> *Daily Telegraph*, 3 June 2000

To see how to use control variables to analyse this issue, consider Table 39, which summarises data from the university admissions scheme, UCAS. This table yields the percentages in Table 40. Note that the gender of the applicant is being taken as the dependent variable, since although sex is determined at birth, the university application happens many years afterwards and might well be determined by the occupation of the head of the household. This is

Table 39 University applicants by class, UK, 2000

Occupational class of head of household	University applicants		Total
	Male applicants	Female applicants	
Professional/intermediate	74 883	86 687	161 570
Skilled/unskilled	50 796	62 945	113 741
Total	125 679	149 632	275 311

Source: adapted from UCAS website

Table 40 University applicants by class and gender, UK, 2000[3]

Occupational class of head of household	University applicants		Total
	Male applicants	Female applicants	
Professional/intermediate	46.3%	53.7%	161 570
Skilled/unskilled	44.7%	55.3%	113 741
Total	45.6%	54.4%	275 311

Source: adapted from UCAS website

a familiar 2 by 2 table. You can see that class makes a 1.6 per cent difference to the gender split of university applications. However, introducing a third variable makes it necessary to split up the table (see Table 41). In this example, the control variable — the acceptance rate — has two categories, accepted applications and rejections. The table therefore splits in two (Table 42), and each cell in our original table (Table 41) splits according to the frequencies of the categories of the control variable. If you add up the corresponding cells in both parts of Table 42 — top left to top left, top right to top right, and so on — you will arrive back at the original Table 41.

Having split the bivariate table, you then analyse each new 'child' table just like before. You could use the new measure V, but let's stick to epsilon for now. Since there are two tables, you get two results from Table 43, 14.6 per cent for accepted applications and −56.3 per cent for rejections. These results are then compared. So here 14 compares with −56, a difference of 70 per cent. This comparison is the contribution of the control variable. If the control variable made no difference, then the two values of epsilon would

Table 41 University applicants by class and gender, UK, 2000

	University applicants		
Occupational class of head of household	Male applicants	Female applicants	Total
Professional/intermediate	74 883	86 687	161 570
Skilled/unskilled	50 796	62 945	113 741
Total	125 679	149 632	275 311

Source: adapted from UCAS website

	Accepted entries			Rejected entries		
Occupational of head of household	Male applicants	Female applicants	Total	Male applicants	Female applicants	Total
Professional/intermediate	57 783	84 233	142 016	17 100	2 454	19 554
Skilled/unskilled	35 200	28 501	63 701	15 596	34 444	50 040
Total	92 983	112 734	205 717	32 696	36 898	69 594

Source: adapted from UCAS website

Table 42 University applications by class, gender and rejection rate, UK, 2000

Table 43 University applications by class, gender and rejection rate, UK, 2000

	Accepted entries			Rejected entries		
Occupation of head of household	Male applicants	Female applicants	Total	Male applicants	Female applicants	Total
Professional/intermediate	40.7%	59.3%	142 016	87.5%	12.5%	19 554
Skilled/unskilled	55.3%	44.7%	63 701	31.2%	68.8%	50 040
Total	45.2%	54.8%	205 717	47.0%	53.0%	69 594

Source: adapted from UCAS website

be the same. If the difference between the two values of epsilon is less than the original value in the bivariate analysis, then the contribution of the control variable is less than that of the independent variable. If on the other hand the difference is larger, the control variable is causing the action. That is the case here. When you look at the original bivariate association there is little difference between the category percentages. When you add the third variable, large differences suddenly begin to appear. If the numbers grow larger, the control variable is where the action is; if the numbers get smaller, the action is in the independent variable. In this example, the acceptance rate is the main cause of differences in university admissions.

Note that since you are comparing two results the signs do matter now – you need to be consistent in the direction of subtraction. So given an epsilon value of 20 per cent in one sub-table and −20 per cent in the other, the difference is 40 per cent, not zero. Similarly, the difference between the values of epsilons in Table 43 is 70 per cent, not 42 per cent. This shows that the associations are moving in opposite directions. A higher proportion of women from professional and intermediate households have their applications accepted, whereas the proportion of male applicants from these households whose applications are rejected is very large. The situation is reversed for applicants from other households. It is the acceptance rates that are making the biggest difference in our example.[4]

Note that the reliability of each separate table still needs to be checked as before. In fact, since the table is split, the numbers are smaller, so the question of reliability is even more crucial. Pay particular attention to the possibility of expected cell counts of less than 5.

HOW TO SIMPLIFY TABLES

While the idea of control variables is straightforward enough, the models that are associated with them can become complex, and communicating the data can become difficult. Briefly, two techniques can be employed for this:

1 *Path analysis*, where the focus is on the effect of the control variables.

2 *Standardised tables*, where the focus is on the original two variables, and the effect of control variables is removed.

This book will focus on the standardisation of tables, which is a simpler technique. The book only mentions path analysis briefly in Chapter 10.

Standardised tables

- **Title** Standardised tables
- **What are they?** A method of recalculating a two-way table to remove the effect of other variables
- **Why use them?** 1 They explain the actual relationship between the variables

    2 They simplify the presentation and understanding of complex tables

The main reason for concentrating on this technique is that it is a very straightforward way of simplifying complex data. It does not focus on exactly the same problem as path analysis, which is concerned with identifying the contributions of each variable to a causal chain. Standardisation in a sense does the opposite, in that it removes the effect of the control variables, leaving us with a simple two-way table to analyse.

Table 44 adds a fourth variable to the analysis of university applications, the variable of race. This gives a four-way table, a 2 by 2 by 2 by 2 table. This table now has four sub-tables, since each of the two sub-tables in Table 42 is now split into two, one for each category of the new variable. If the new variable had three categories, there would be six sub-tables.

As before, each of the sub-tables is analysed in the usual manner by percentaging the independent marginals. This yields Table 45 (ignore the numbers in parentheses for now).

Table 44 University applications by class, gender, rejection rate and race, UK, 2000

	Accepted applications			Rejected applications		
	Male	Female	Total	Male	Female	Total
White applicants						
Professional	56 405	81 326	137 731	16 333	2 416	18 749
Skilled	33 778	27 364	61 142	14 617	32 170	46 787
Total	90 183	108 690	198 873	30 950	34 586	65 536
Black applicants						
Professional	1 378	2 907	4 285	767	38	805
Skilled	1 422	1 137	2 559	979	2 274	3 253
Total	2 800	4 044	6 844	1 746	2 312	4 058

Source: adapted from UCAS website

Table 45 University applications by class, gender, rejection rate and race, UK, 2000

	Accepted applications			Rejected applications		
	Male	Female	Total	Male	Female	Total
White applicants						
Professional	41.0%	59.0%	137 731	87.1%	12.9%	18 749
Skilled	55.2%	44.8%	61 142	31.2%	68.8%	46 787
Total	45.3%	54.7%	198 873	47.2%	52.8%	65 536
	$\varepsilon =$	14.3%	(0.722)	$\varepsilon =$	-55.9%	(0.238)
						$d\varepsilon = 70.2\%$
Black applicants						
Professional	32.2%	67.8%	4 285	95.3%	4.7%	805
Skilled	55.6%	44.4%	2 559	30.1%	69.9%	3 253
Total	40.9%	59.1%	6 844	43.0%	57.0%	4 058
	$\varepsilon =$	23.4%	(0.025)	$\varepsilon =$	-65.2%	(0.015)
						275 311 = N
						$d\varepsilon = 88.6\%$

Source: adapted from UCAS website

While this book does not emphasise path analysis, it is useful to compare the two approaches briefly. A path analysis approach would continually calculate and compare values of epsilon. So it would take the same calculation for the difference in values of epsilons as in Table 43, one for each of the categories of the control variable. Let us refer to these as dε, for 'difference in epsilon' (I'm no better at the naming game than anyone else). There are two categories in Table 45, so you get two values for the difference in epsilon, two values of dε. These are then compared. The difference between them is the contribution of the new variable. Again, if the numbers are growing bigger, the new variable is where the action is. If they're growing smaller, the other variables have the bigger effect: they make the most difference to the outcome. In Table 45, the values of epsilon for the white applicants were 14.3 per cent and −55.9 per cent. This is a difference of 70.2 per cent. The figures for black applicants were 23.4 and −65.2 per cent, yielding a difference of 88.6 per cent. If the new variable, race, had no effect these differences would be the same. The more they vary the bigger the effect race can be said to have. In this example, it can be said to have a 18.4 per cent effect, the gap between 70.2 per cent and 88.6 per cent. This is less than the effect of acceptance rates, however, since the results are getting smaller.

However, this analysis is a bit difficult to understand, and it is just as hard to communicate. Tables 44 and 45 are also quite large, and they have too many numbers to really be readily understandable. The technique of standardisation is an alternative method. The idea is to take this large table and combine the cells so as to arrive back at a two-way table. However, this new table will remove the effect of the control variables.

The first step in the logic is to figure out the proportion[5] of the total that each table represents ($N = 27,533$). So the top left sub-table (white, accepted applications) in Table 45 makes a contribution of $198,873/27,533 = 0.722$, or just under three-quarters. Rejected white applicants account for just under one-quarter. The proportion of each sub-table is calculated in a similar way, and these are the numbers in parentheses in Table 45. The proportions of all the sub-tables will sum to 1.

HOW TO SEE IF SEVERAL THINGS ARE LINKED· 69

Table 46 University applications by class and gender, standardised for rejection rate and race, UK, 2000

	Male	Female	Total = 100%
Professional	52.5%	47.5%	161 570
Skilled	49.2%	50.8%	113 741
Total	45.6%	54.4%	275 311

41 × 0.722 = 29.6%
87.1 × 0.238 = 20.7%
32.2 × 0.024 = 0.8%
95.3 × 0.014 = 1.4%
Total = 52.5%

These same as original bivariate table

Source: adapted from UCAS website

The proportion of each sub-table is then multiplied by the percentage in each of its cells to arrive at a new adjusted percentage. These new figures are then added to reconstruct the original 2 by 2 table – all the top lefts, all the top rights, and so on. This gives Table 46, which shows the association between the original variables with the effect of the control variables removed. So in Table 46, gender has a 3.3 per cent effect on class when the effects of race and rejection rates are removed. Of course, you could conduct a similar table for those variables and thereby find out which has the largest effect. The table is short and simple to understand.

While the example here was limited to 2 by 2 tables, the technique can also be used on larger tables. In addition, standardisation has the effect of restoring the original sample size, which makes the results more reliable.

SUMMARY: MULTIVARIATE ANALYSIS

- This section considered models with more than two variables. The extra variables may be prior to or intervening between an independent and a dependent variable. These are called control variables.
- Theoretical analysis is needed to identify possible control variables.
- To analyse control variables, a two-way table is a split and the results compared. The bigger the difference in results, the bigger the effect of the control variable.
- When analysing a large number of control variables simultaneously, or for control variables with many categories, standardisation can be used.[6]

APPENDIX

Calculating Cramer's V

This is fairly straightforward.

1 Count the number of categories on the shortest side of the table (i.e. the lesser of the number of rows or the number of columns). Subtract 1 from this. So for a table of four rows and three columns, this would be 2. For a 2 by 2 table or a 2 by 3 table this is 1.

2 Multiply this by N.

3 Divide chi-square (χ^2) by the result of (2).

4 Calculate the square root of (3).

$$V = \sqrt{\chi^2 / N \times (\text{short side} - 1)}$$

So for the example in table 37 page 59:

1 The 'short side' is 1 (since it's a symmetrical table, you can use either rows or columns).

2 N is 50,912. 50,912 × 1 is still 50,912.

3 chi-square is 11,430. 11,430 divided by 50,912 is 0.224.

4 The square root of 0.224 is 0.473, or 47.3 per cent.

5 So $V = 47.3$ per cent.

Notes

1 While these are disadvantages from a practical viewpoint, it could be suggested that these constraints are not wholly evil. Ascertaining the direction of association, for example, might help in understanding the causal paths and in building a causal model or a getting a theoretical perspective, however casual. It is also a moot philosophical question as to what an association between three or more nominal categories actually *means*. However, these questions are beyond the scope of this book, which is only a practical introduction to data analysis.

2 See the appendix at the end of this chapter.

3 Tables 39 and 42 are significant at the 1 per cent level (they're 99 per cent reliable).

4 It is worth reiterating here that these examples are only to illustrate the analysis techniques and should not be used for policy analysis. I have taken many liberties with the data in this example.

5 This is one of the few times when I find it easier to use and talk about decimals rather than percentages.

6 As was stated, the alternative approach – path analysis – is beyond the scope of this book.

Chapter 6

HOW TO SEE IF PAIRS ARE LINKED

- **Name** Pearson's correlation
- **What it does** It shows if two sets of magnitudes (numbers) are associated and in what direction
- **Why use it?** 1 Most powerful and accurate of the measures of correlation

    2 Gives reliable results for small samples

So far this book has looked at association at the nominal and ordinal level: association between variables that are either only categories or can be placed in a 'league table' order. This chapter will look at correlation at the ratio level: correlation between magnitudes (what most people would call 'numbers'). These methods look to see if the actual values of two variables are associated with each other.

As with the rank correlation applied to league tables, you can check both the *strength* and the *direction* of the association. Like rank correlation, if the numbers change in the same direction the correlation is positive; if they change in opposite directions, this is called negative correlation.

SCATTERGRAPHS

The simplest method of looking at any correlation is to use a graph, such as Figure 10. This plots the birth rate of a randomly selected

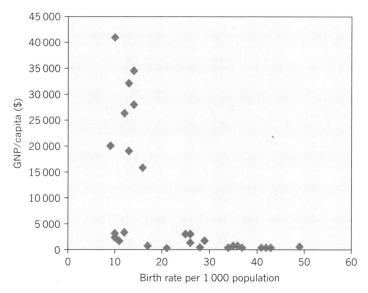

Figure 10 GNP per capita by birth rate
Source: adapted from *World Guide 1999/2000*

sample of countries against their wealth (GNP per capita). Figure 10 is called a *scattergraph*. The closer the points on the graph are to a straight line the stronger the correlation. It seems to the eye that there is indeed a fairly strong correlation in Figure 10 between wealth (or poverty) and birth rates. The argument is that 'the rich countries get richer, and the poor countries get babies'.

Graphs are a good way of understanding the logic of correlation. Consider Figure 11. This is an imaginary scattergraph with the points fitting roughly inside an oval shape. This graph is unusual and is different to Figure 10 in that the vertical and horizontal axes are not zero. They are actually the means of the two variables. They will meet in the centre of the oval, since that is what means show; they measure the centre of a group of data.

This divides the graph into four quadrants. The top right has data that is above average for both the variables. The points in the bottom left quadrant are below average for both variables. The other

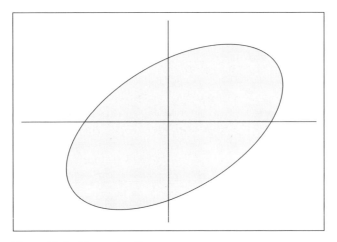

Figure 11 Positive oval scattergram

two quadrants are below average for one variable but above average for the other.

You may recall your elementary mathematics, when you will have learned that multiplying two positive numbers together produces a positive product, as does multiplying two negative numbers together. However, if you multiply two numbers of different sign, the result is negative. This idea can be applied here. In the top right quadrant the area under all the points is positive, since the data is positive (above average) on both axes, so you are multiplying positive by positive. The area in the bottom left quadrant is also positive, since the points are negative on both axes, so you are multiplying negative by negative. In the other two quadrants the points have opposite signs, and the area between points and axes will be negative.

Now consider what happens when the oval is squeezed, as in Figure 12. The proportion of points in the top right and bottom left quadrants – the positive quadrants – rises, and the proportion in the negative quadrants falls. In fact, the more the oval is squeezed (Figure 13), the more the proportion of positive to negative areas increases, since the two means will always be in the centre of the data.

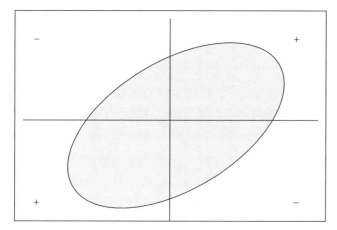

Figure 12 Sign of quadrants

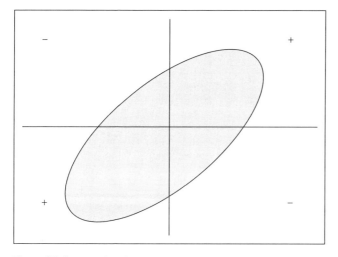

Figure 13 Squeezed oval

In Figure 14, very little of the oval is in the negative quadrants. Ultimately, the data would be in a straight line pointing north-east (Figure 15(a)). This would be a correlation of +100 per cent. A straight line pointing north-west would represent a correlation of −100 per cent (Figure 15(b)). The same logic would hold for an oval

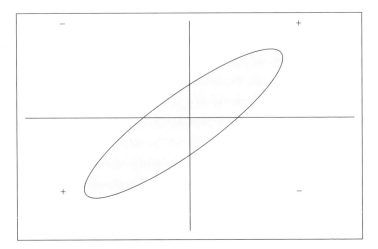

Figure 14 Flat oval

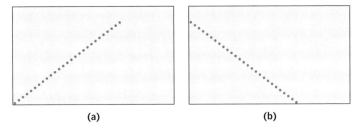

Figure 15 Perfect correlations: (a) perfect positive correlation; (b) perfect negative correlation

with axis sloping up from left to right. If the data was in a circle rather than an oval (Figure 16), this would mean a correlation of zero, since there would be as many points in negative as in positive quandrants.

All that the correlation measures is whether there is a pattern to the variations from the mean: that is, whether the dispersions you looked at in Chapter 1 form a regular pattern or not. It compares the dispersions in both directions: whether the variation from the mean in one variable matches that of the other variable.

The standard measure for correlation at the ratio level was devised by a chap called Pearson and is called the *product moment* correlation.

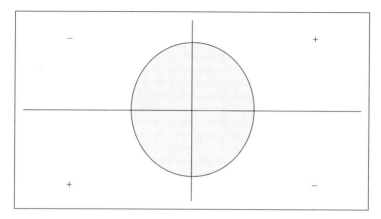

Figure 16 Zero correlation

It is given the letter r. Like the rank correlation looked at in Chapter 4, Pearson's correlation gives a percentage, which tells the strength of association.[1] Returning to the population question, the data producing Figure 10 is presented in Table 47. This gives an r of minus

Table 47 GNP per capita by country and birth rate, 1996

Country	Birth rate (per 1 000 population)	GNP/cap US$	Country	Birth rate (per 1 000 population)	GNP/cap US$
Japan	10	40 940	Romania	11	1 600
Norway	14	34 510	Morocco	26	1 290
Denmark	13	32 100	China	17	750
USA	14	28 020	Honduras	35	660
France	12	26 270	Lesotho	36	660
Italy	9	19 880	Guinea	49	560
Canada	13	19 020	Mongolia	28	360
New Zealand	16	15 720	Kenya	37	320
Slovakia	12	3 410	Haiti	34	310
Estonia	10	3 080	Togo	43	300
Venezuela	26	3 020	Angola	49	270
Lebanon	25	2 970	Guinea-Bissau	41	250
Russia	10	2 410	Madagascar	42	250
Latvia	10	2 300	North Korea	21	250
El Salvador	29	1 700	Tanzania	42	170

Source: adapted from *World Guide 1999/2000*

59.7 per cent, confirming a fairly high correlation between birth rate and the wealth of nations. The correlation is negative, because GNP decreases as the birth rate increases.[2]

Reliability

To consider the reliability of the result, a similar table to that used for r_s is utilised. Bear in mind that, just as with nominal and ordinal association, you want to find out about the population, not the sample. The table is reproduced at the end of the chapter. The higher both r and sample size N, the more reliable our data is likely to be. In our example, with a sample size of 30, the data is significant at the 1 per cent level; it is 99 per cent reliable.

Note how the level of reliability for fairly small samples is higher than for rank correlation. This is because not only is the order of rankings matching up, but the actual magnitudes are also matched. This is one advantage in using the higher-level correlation techniques, since the reliability of the results are higher.

One-tailed tests

This seems a good place to introduce a method for improving the reliability of the measures. If you can specify not only that there is a correlation but also in which direction that relationship will be, then the reliability of the correlation can be doubled. So a correlation that was significant only at the 10 per cent level (90 per cent reliable) is now reliable at the 5 per cent level (95 per cent reliable). Similarly, a correlation previously reliable at the 2 per cent level is now reliable at the 1 per cent level (this is, incidentally, one reason why statisticians do the logic 'backwards'). This is called a 'one-tailed' test, for reasons discussed later, as opposed to the non-directional two-tailed tests used in previous chapters.[3]

So, in the example, suppose that you had data for only the first ten countries listed in Table 47, but still had the same correlation, −56.5 per cent. The hypothesis you are going to test is that GNP and birth rate will move in opposite directions. If GNP rises as the birth rate rises, this hypothesis will not be supported anyway, no matter how strong the correlation or how reliable the data. A 56.5 per cent correlation and an N of 10 would not be reliable at the 5 per cent level in the non-directional test you did before. But since you are now specifying the direction of the correlation, the chances of this correlation being due to luck are halved, and the result is now reliable (statistically significant) at the 5 per cent level. You can now generalise from this result to the rest of the population: the other countries of the world.

By the same token, consider the argument in Chapter 3 that the mortality index in UK hospitals falls as the doctor/patient ratio falls. It would be bizarre if the mortality index rose the more doctors there were, so a reasonable hypothesis is that the lower the ratio the lower the index, that there will be a positive correlation. So you would test not only to see if the teaching and research index numbers are associated but also specifically to see if the numbers change in the same direction: a positive correlation. A negative correlation would refute the theory anyway, so there would be no need to test reliability on this.

Recall that the rank correlation coefficient (r_s) for the data was 60 per cent. Since N is 18, this would be reliable only at the 5 per cent level on a two-tailed test, but because you specified the direction of the association you can use a one-tailed test and increase the reliability to 1 per cent.

Tables for one-tailed tests are given in the appendix.

LIMITATIONS OF LINEAR CORRELATIONS

The techniques for rank and product moment correlation that have been outlined are suitable only for 'straight line' correlation, correlation where the pattern follows a single direction. It is entirely possible for the data to have a pattern that is not a straight line, which is why it is always a good idea to produce a scattergraph of rank or product moment correlations and eyeball it. For example,

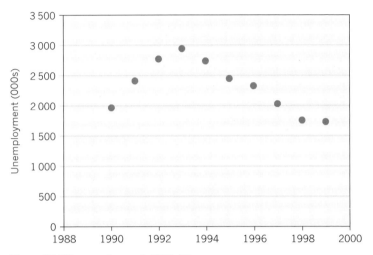

Figure 17 UK unemployment, 1990–99

later in this chapter I will show how to use time as a variable and consider time series, but the same correlation analysis can be used. The unemployment trend over the last decade is plotted in Figure 17. Clearly something is happening, since there is a clear pattern, but it is not a straight-line correlation.

Analysing such a non-linear relationship requires advanced techniques, which are beyond the scope of this book. Broadly speaking, they involve changing the number scale so that the graph become a straight line.[4] In the example, you might change the units with which you measure unemployment.

Another limitation of Pearson's *r* is that it cannot be used on very skewed data, or on data that does not follow a 'normal' pattern of distribution. There is more on normal distributions in the next chapter, but this is a good reason why rank correlations may sometimes be preferred, since there is no question of 'normal' distribution in league tables. The technical term for this restriction on data spread is a *parameter*, and tests that are dependent on that spread of data are called *parametric* tests. Spearman's rho, on the other hand, is a *non-parametric* test and makes no assumptions about the spread of data.

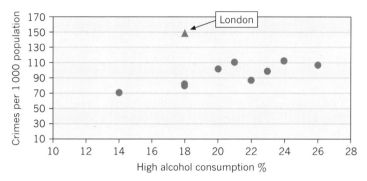

Figure 18 Crimes committed by men aged over 16 by region in relation to high alcohol consumption, Great Britain, 1999

A final limitation on the use of correlation tests is that they are affected by 'odd' results, which are called *outliers*. Pearson's *r* is more susceptible to the effect of outliers. Consider Figure 18, which is a scattergraph of the percentage of men in each English region who drink over eight units of alcohol a day against the crime rate in that region. This data gives a low correlation of 32 per cent, since although there is a clear pattern it is influenced by a single outlier marked by a triangle – in this example the London region. There may be a case for removing such outliers, since they distract from a very clear data pattern. Such a process is called 'data smoothing' and is as much art as science, and indeed whether such doctoring of the data is permissible is always open to debate. In this example, removing or 'correcting for' the outliers increases Pearson's *r* from 32 to 82 per cent.

SUMMARY

- This section dealt with association at the ratio level.
- Ratio-level correlation can be positive or negative, just like rank correlation. Data in positive correlations changes in the same direction; negative correlation data changes in the opposite direction.

- The standard measure for ratio-level correlation is Pearson's *r*. This measures the correlation between actual numbers, not just their ranking.
- Pearson's *r* gives higher levels of reliability than Spearman's rho. However, it does make assumptions about the spread of data.
- Both measures of correlation, product moment and rank correlation, can be made more reliable by specifying the direction of the association.
- Both measures consider only straight-line associations. The data may have a pattern that is not a straight line. It is best to check on a scattergraph.

PREDICTING ONE VALUE FROM ANOTHER

- **Name** Linear regression

- **What it does** It enables the prediction of the value of one variable given the value of another associated variable

- **Why use it?** Prediction is useful for decision making. Also used to analyse trends over a period of time

This section takes the previous techniques one stage further and looks at how to actually predict a value for ratio variables. This is because at the ratio level not only can you investigate the possibility of an association but you can also measure what that association actually is. That is, you can measure one variable given the value of another.

Consider the scattergraph in Figure 19, which is Figure 18 with the outlier (London region) removed. This, it will be recalled, has a correlation of 82 per cent. What you are trying to do here is similar to what you were doing in Chapter 1. Instead of getting one number to describe the divergence from the mean, here you try to draw a single straight line that describes the association between the two variables. The logic behind this is similar to the logic of standard

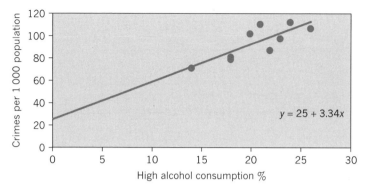

Figure 19 Crimes committed by men aged over 16 by region in relation to high alcohol consumption, Great Britain, 1999

deviations. No single straight line will join every data point in the graph, just as no mean equals every number in a set. What you are aiming to find is a line that minimises the gaps from the points. Here this is taken as meaning the lowest total differences squared (just as the standard deviation was the sum of squared differences). Hence the method is called *least squares regression*. It is the 'line of best fit' through the data.

The higher the correlation, the closer the points will be to this line and the better a predictor the line will be, just as a low standard deviation shows that the mean is a good description. The graph of Figure 19 clearly has its points close to the regression line, which is to be expected with such a high correlation. For this reason, it may be a good idea to include a least squares line in a scattergraph, because it acts as a visual aid.

The regression line has two components. The first is an indication of how far up the graph the line is. This is called the *intercept*, because it is where the line would cut the vertical axis if extended (the vertical axis is usually called, in a burst of originality, the y axis). The intercept is the value of y when the horizontal axis (called, predictably, x) is zero. To continue the naming game, the intercept is called 'a'. In Figure 20, the larger a is, the higher up the y axis the line starts. The second component of the regression line is the *slope*. This shows how far around the intercept the line 'pivots'. In Figure 21, the lower the number the more the line 'pivots'

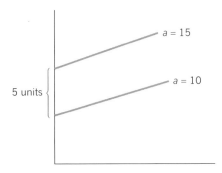

Figure 20 Intercepts

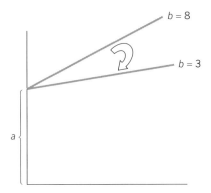

Figure 21 Slopes

around. The slope of the line is called (wait for it) '*b*'. Many readers will already be familiar with this idea, since it is in the Highway Code (Figure 22). The higher the number, the steeper the slope.[5]

The regression line is therefore described as $y = a + bx$, where:

- *a*   is a constant number (the intercept)
- *b*   is the slope
- *x*   is the variable
- *y*   is what you are trying to find out.

In Figure 19 the intercept is 25, and the slope is 3.34, or about $3\frac{1}{3}$. The crime rate is $3\frac{1}{3}$ times the proportion of heavy drinkers plus

Figure 22 Road sign indicating steepness of slope

25. So, given a proportion of heavy drinkers of 21 per cent (the national average), the predicted crime rate is 95.14 per thousand of population.

A north-west slope is, you may recall, positive and a north-east slope negative. So if b is positive the line slopes north-west, and if b is negative it slopes north-east.

In Figure 23, the bottom line has a lower intercept (a) value, which is why it is the lower line. It has a negative slope (b value) and so slants south-east. The upper line has a higher intercept and a positive slope, so it slants north-east. Once you have the formula for these lines, you can pump in any value for x, and the prediction for y comes up. So for Figure 19, if say, 30 per cent of adult men drink over eight units a night, this would give a crime rate of 125.2 per thousand of the population.

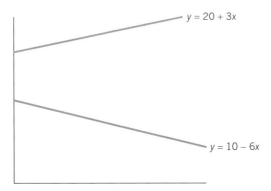

Figure 23 Positive and negative slopes

Trends

One use of regression analysis is to analyse trends – how variables change over time. Time can be used as a variable – it is a continuous, ratio variable. You can plot a trend just the same as for any other graph, with time as the independent variable.[6]

Consider, for example, the question of the number of senior citizens that the UK will have in the future. This is a vital issue, since if the numbers increase the costs to the welfare state also increase, as well as affecting pension payments. This means that other welfare expenditure, and the taxes needed to finance them, will also be affected.

BOX 5 PENSION TIME BOMB?

Put more in your pension now to avoid crisis later

Experts have criticised the pensions industry, government and managers of occupational pension schemes for failing to warn the public of an impending crisis.

By 2008, for the first time in history, there will be more retired people in Britain than children, with the ratio of pensioners to workers increasing every year.

This demographic shift is set to place an unbearable strain on state pensions and render any government promises about future state payouts meaningless for today's younger savers.

Experts say the 'short termism' of successive governments has meant that none has wanted to draw attention to the failings of the system, or the true cost of adequate retirement savings.

This demographic timebomb – caused by a falling birth rate, rapidly growing pensioner population and declining stock-market returns – will force people to contribute more to their pensions. Investment returns are expected to fall because of lower inflation and interest rates. And it is forcing companies to change their generous occupational final-salary schemes to less advantageous money-purchase plans.

The seemingly innocuous change could cost workers several times as much in contributions to achieve the same retirement income.

The Sunday Times, 11 June 2000

> This report of an ever-increasing burden of pensioners was, incidentally, rejected by a royal commission:
>
> ### Care shake-up for elderly urged
>
> Report dismisses fears of demographic timebomb
>
> Arrangements for long-term care of more than 1 million elderly people are sustainable neither morally nor practically, Sir Stewart Sutherland said yesterday as he presented the report of the first royal commission in the field of health for more than 20 years.
>
> The report, *With Respect to Old Age*, sets out a model for revamping long-term care in the medium run, redefining the divide between what is state-funded and what is means-tested at an initial cost of about £1 billion.
>
> The commission has concluded that while the population is ageing, there is no 'demographic timebomb' threatening unaffordable care costs. It has rejected the idea of looking to private insurance against such costs and decided that it would be unfair to adopt a funded social insurance scheme.
>
> *The Guardian*, 2 March 1999

Regression analysis can be used to investigate this question. Figure 24 plots the trend of the proportion of senior citizens since 1951. You may be used to seeing these points joined up, but to predict future values you can use a least squares regression line instead. The points in Figure 24 fit quite closely around the line to give a correlation of 98 per cent. The line gives a formula of $y = 0.115x - 211$. Note that these figures appear a bit different to those before. First, the intercept is a minus figure, since the data starts in 1951, not AD 1. For this reason, the figure has been put at the end of the equation, since it is less confusing that way. Also note that the graph is scaled in percentages, so the answers will be percentages (the slope 0.115 is 11.5 per cent, but typing in 0.115 is easier). The equation confirms the very shallow slope that can be seen in the graph.

So to predict the level in 2001, multiply 2,001 by 0.115 and subtract 211, which gives a prediction of 19.1 per cent. This book

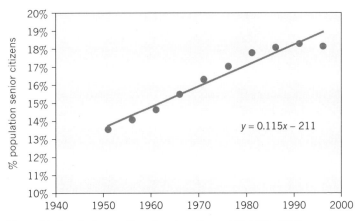

Figure 24 Trend of proportion of senior citizens, UK, 1951–1996

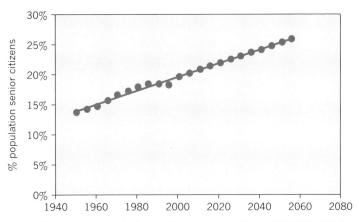

Figure 25 Projected trend of proportion of senior citizens, UK, 1951–1996; projection to 2056

was written before the 2001 census, so it will be interesting to see if the trend continues. Of course, the real interest in this example is to predict into the future. Figure 25 is what the trend would look like for the next fifty years. This suggests that if the trend of the past fifty years continues, just over one-quarter of the UK population will consist of senior citizens, compared with about one-fifth at present.[7]

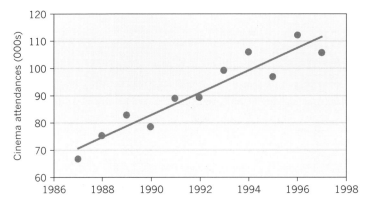

Figure 26 Cinema attendances, UK, 1987–1997

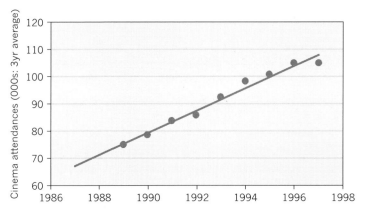

Figure 27 Cinema attendances, three-year moving averages, UK, 1987–1997

Time series can have some peculiarities, and there are many techniques to help with their analysis. One simple technique is to replace the data with a *moving average*. Here each data point is replaced with that average over, say, the last three years. This helps to smooth out any fluctuations in the trend. The fluctuations around the trend line in Figure 26 have been smoothed out in Figure 27 by plotting a moving average over three years. This smooths out the data and makes the underlying trend easy to follow. Again, this is an art as much as a science. Different averages give different results. However, the more time periods in the moving average the greater the smoothing effect.

There are more complex variations on the moving average idea. One of note is *exponential smoothing*, where the averages are weighted according to past predictive accuracy, and more recent data is also weighted more highly.

Limitations

Finally, bear in mind that this technique is valid only for linear (straight-line) regression. If the association follows a different pattern, more advanced techniques must be used. Bear in mind that all the limitations for using linear correlation also apply to linear regression.

SUMMARY

- This section was concerned with predicting one variable from another.
- Prediction involves describing the correlations. On a scattergraph, it is represented by a line of best fit. The closer to the line the points are, the higher the correlation and the stronger the prediction.
- The line of best fit can be represented as a formula. This enables the prediction of values of the dependent variable.
- Time can be used as a variable and regression techniques used to analyse trends.
- Techniques such as moving averages can be used to smooth out data.

APPENDIX

Table 48 Minimum values of Pearson's r for reliability

	Non-directional (two-tailed) tests		Directional (one-tailed) tests	
	Level of reliability		Level of reliability	
N	5% level (95% certainty)	1% level (99% certainty)	5% level (95% certainty)	1% level (99% certainty)
4	95.0%	99.0%	90.0%	98.0%
5	87.8%	95.9%	80.5%	93.4%
6	81.1%	91.7%	72.9%	88.2%
7	75.4%	87.5%	66.9%	83.3%
8	70.7%	83.4%	62.2%	78.9%
9	66.6%	79.8%	58.2%	75.0%
10	63.2%	76.5%	54.9%	71.6%
11	60.2%	73.5%	52.1%	68.6%
12	57.6%	70.8%	49.7%	65.9%
13	55.3%	68.4%	47.6%	63.4%
14	53.2%	66.1%	45.7%	62.1%
15	51.4%	64.1%	44.1%	59.2%
16	49.7%	62.3%	42.6%	57.4%
17	48.2%	60.6%	41.2%	55.8%
18	46.8%	59.0%	40.0%	54.3%
19	45.6%	57.5%	38.9%	52.9%
20	44.4%	56.1%	37.8%	51.6%
21	43.3%	54.9%	36.9%	50.3%
22	42.3%	53.7%	36.0%	49.2%
27	38.1%	48.7%	32.2%	44.5%
32	34.9%	44.9%	29.6%	40.9%
37	32.5%	41.8%	27.5%	38.1%
42	30.4%	39.9%	25.7%	35.8%
47	28.8%	37.2%	24.3%	33.8%
52	27.3%	35.4%	23.1%	32.2%
62	25.0%	32.5%	21.1%	29.5%
72	23.2%	30.2%	19.5%	27.4%
82	21.7%	28.3%	18.3%	25.7%
92	20.5%	26.7%	17.3%	24.2%
102	19.5%	25.4%	16.4%	23.0%

Source: adapted from R.D. Nelson, *The Penguin Book of Mathematical and Statistical Tables* (Harmondsworth: Penguin, 1980)

Table 49 One-tailed, directional test: minimum values of Spearman's rho for reliability

N	5% level (95% certainty)	1% level (99% certainty)
		Level of reliability
4	100.0%	–
5	90.0%	100%
6	82.9%	94.3%
7	71.4%	89.3%
8	64.3%	83.3%
9	60.0%	78.3%
10	56.4%	73.3%
12	50.3%	67.8%
14	45.6%	64.5%
16	42.5%	60.1%
18	39.9%	56.4%
20	37.7%	53.4%
22	35.9%	50.8%
24	34.3%	48.5%
26	32.9%	46.5%
28	31.7%	44.8%
30	30.5%	43.2%

Source: adapted from R.D. Nelson, *The Penguin Book of Mathematical and Statistical Tables* (Harmondsworth: Penguin, 1980)

Notes

1 Again, most textbooks will report a decimal instead of a percentage.

The method for calculating r can be found in Clegg, F., *Simple Statistics* (Cambridge: Cambridge University Press, 1990). However, there are cheap pocket calculators that will do linear regression, so it seems a bit redundant to recap the calculation method.

2 The question of population and development is fraught. See e.g. Barry, J., *Environment and Social Theory* (London: Routledge, 1999) chapter 7 for more on this issue.

3 Tables for directional one-tailed tests are given in the appendix to this chapter.

4 Common techniques are to use mathematical functions such as square roots and logarithms.

5 The slope will only be a percentage if the units are the same for both variables, however, which they are not in the example in this chapter.

6 Time is always the independent variable; in fact, time is the ultimate independent variable.

7 For arguments as to why this may not occur, and why it may not be a problem, see Mullins, *The Imaginary Time Bomb* (London: Tauris, 1999).

Chapter 7

HOW TO JUDGE THE RELIABILITY OF AVERAGES

- **Title** Confidence intervals

- **What they do** They tell you a range within which the mean of a population is likely to fall

- **Why use them?** They enable you to generalise from the mean of a sample to the mean of a larger population from which the sample is drawn. They can also be used to check if differences between means are 'real' or due to luck

The last few chapters have dealt with the question of association rather than description, and great emphasis was placed on testing each measure for reliability. What the next two chapters will cover is the application of the question of reliability to the descriptions of a group of numbers. The use of the mean and standard deviation were covered in Chapter 1, but in order to aid understanding the idea of reliability was left out. The question of reliability is just as valid for a description of data from a sample as it is for measures of association, however, and this chapter will fill that gap.

When describing a set of numbers, it is just as desirable to know about the population rather than the sample as it is when analysing an association. In particular, it is important to know whether the mean of a sample of ratio-level data (age, income, and so on) accurately describes the population it is drawn from. That is to say, would another sample yield the same or a similar result? If so, how similar would the results be?

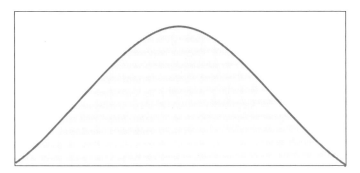

Figure 28 Distribution of sample means

This is the logic behind confidence intervals. While we cannot know for sure what the mean of a large population is short of looking at all of them, we can know what the means of a large number of samples would probably look like.

Consider for instance a UK sample of 1,000 people. Let us assume we're describing their income, which we will also assume is £397.26 a week on average. What you want to know is whether this mean income figure also holds for the rest of the UK population. If you took a second sample to test it, it is not too likely that the mean would be exactly £397.26 a week. It might be close though; it would not be surprising if it was a few pounds or so either side. Similarly, it would be surprising if a third sample had the mean income exactly the same to the nearest penny as either of the other two, but again you might expect it to be reasonably close.

Supposing you took 1,000 samples of 1,000 respondents each, which is 1 million people. This would yield 1,000 means. We can guess pretty well how these would look – they would look like Figure 28, which is a histogram representing the frequencies of the means of each of a large number of samples. Most of these means will be grouped around the centre, with fewer as you move towards the edges of the graph. If you wished, you could use the descriptive techniques from Chapter 2 to describe this new graph. You could use measures of central tendency and measures of dispersion, and so on. In doing this, you should note the following:

1 The new graph is symmetrical. This is to be expected – the variations from the centre of a large number of results would be both higher and lower in equal quantities, and there would be no skew. That is, the mean, median and mode would be the same and would be the highest point on the graph.

2 While we do not know the mean of the population for sure, if you take a huge number of samples and average out the means from them, this will be the same as the population mean. The larger the number of samples, the closer the average of their means will be to the population mean. (This is called the *central limit theorem*.) A good way of thinking about this is to imagine ten coin flips. There may be five tails, but there may easily be four or six, and occasionally more or less than this. However, if you take a large number of sets of ten flips of the coin, the most frequent number of tails will be five, with the frequency falling the further away you get either side, and the *average* of the sets will be five tails.

Moving on to the dispersion of our new (hypothetical) data, you would expect the standard deviation to be less than the standard deviation of the actual data in any of the samples (the new data is a set of means, after all). In order to avoid confusion, the dispersion of the results in a hypothetical set of means is called the *standard error*[1] to differentiate it from the standard deviation of a sample, although it is the same dispersion measure. There is a relationship between the standard deviation in a sample and the standard error of a set of means:

standard error = standard deviation/$\sqrt{N}$

So the bigger the sample size (recall that N is the sample size) the smaller the standard error, and the standard error in the population is much less than the standard deviation in a sample.

The curve in Figure 28 is called the *normal curve*. The properties of the normal curve – and the pay-off for all this – are that you can predict that 95 per cent of the dots in Figure 28 will be within about two standard errors of the mean (Figure 29). Furthermore, 99 per cent of the dots will be within about $2\frac{1}{2}$ standard errors of the mean (Figure 30).

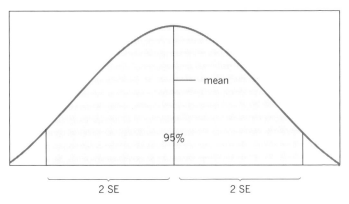

Figure 29 Distribution of normal curve

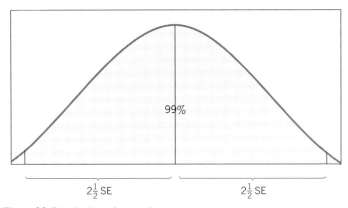

Figure 30 Distribution of normal curve

These match the levels of reliability you have been using, and they can be used to get limits for the population mean. You cannot say for sure what the population mean is, but you can be reasonably confident that it falls within certain limits, which is the pay-off for all this theory. Taking the sample mean as the 'best guess', you can be 95 per cent certain that the population mean is within about two standard errors of the sample mean and 99 per cent certain that it is within about $2\frac{1}{2}$ standard errors of the sample mean.[2] These are called *confidence intervals*.

Calculating Confidence Intervals

1 Obtain the mean and standard deviation of the sample.

2 Calculate the standard error (standard error = standard deviation/ $\sqrt{N}$).

3 Choose a level of reliability: usually 95 or 99 per cent.

4 The population mean is:
   - 95 per cent certain to be within two standard errors of sample mean.
   - 99 per cent certain to be within $2\frac{1}{2}$ standard errors of sample mean.

Returning to the example in Chapter 1 concerning alcohol consumption (Table 50), the mean weekly consumption of 6,342 men in the sample is 13.06 units, with a standard deviation of 11.26 units, a wide variation of about 86 per cent. The standard error for this example is 0.14 units of alcohol.[3] Two standard errors is 0.28, or just over a quarter of a unit. You can be 95 per cent certain that the average alcohol intake of the male population is within 0.28 units each side of the mean. That is, the population mean is between 12.78 and 13.34 units.

Because the sample is large, the confidence limits are narrow. Note how if the sample size was only fifty the limits would be much larger. The standard error would then be 1.59 units,[4] and two standard errors would then be 3.18 units. You could only be 95 per cent confident that the mean is between 11.47 and 14.65 units, a much wider spread.

Table 50 Alcohol consumption by adult men in the UK

Alcohol units consumed	Frequency
0	469
<1	514
1–10	2 325
11–21	1 339
22–40	1 695
Total	6 342

Source: adapted from *Living in Britain 1998*, HMSO (1999)

Confidence intervals for a proportion

When using category data the mean is not used as a descriptor. Instead, you can describe the proportions in each category. So, for example, according to the 1998 General Household Survey 33 per cent of women working in unskilled manual jobs smoke.

You can obtain standard errors and confidence intervals for these proportions in a similar manner to the method for means,[5] but the formula is a bit more complex:

$$\text{standard error} = \sqrt{\%\text{age} \times (100 - \%\text{age})/N}$$

So if 33 per cent of women smoke, and given the sample size of 7,830, the standard error would be 0.53 per cent, or just over half of 1 per cent.[6] Two standard errors is 1.06 per cent. This is the minimum margin of error. You can be 95 per cent certain that between 32 and 34 per cent of unskilled manual women workers smoke. This compares with between 13.7 and 14.3 per cent of professional women, but comparing two proportions is the subject of the next section.

Checking the difference between two means

One of the other pay-offs of being able to check the reliability of a mean is to use the same technique to check on a difference between means. Much policy science revolves around perceived differences between groups – such as different income or employment prospects between genders or races, racial differences in crime rates, differences in the health of different social classes, and so on. It is useful to be able to obtain averages and percentages for different groups and then compare them.

Let us take the example of earnings. According to the Office of National Statistics, the average (mean) hourly rate of pay for men is about £9.50, compared with £7.50 for women. The implication is that discrimination may exist in the employment market.

The issue to be raised in this section is as follows. If you obtain a mean from any two samples, in this case men and women, you would not expect them to be exactly the same even if both populations had the same mean. You wouldn't even expect the same mean if you drew two samples from one population, such as two samples of women non-manual workers. The chances are that there would be *some* difference, even if only a few pennies. The point is that if the mean income of the female group is almost bound to be different from the male group anyway, how can the question of unequal pay be investigated? At what point do you say that the pay is unequal?

> **BOX 6 EQUAL PAY?**
>
> Ministers urged to review women's pay
>
> The government will today be challenged to review women's pay in the public sector as a hard-hitting report finds that Britain's gender pay gap is the worst in the European Union. Thirty years after the Equal Pay Act, women in full-time work are still earning only four-fifths of what men do.
>
> The Equal Opportunities Commission's equal pay task force will press ministers to compel employers to conduct pay audits to address the gender gap. Its report, *Just Pay*, will highlight Britain's place at the bottom of the EU league on equal pay. While women working full-time earn on average 82% of male full-time hourly earnings – up from 69% in 1971 – part-time female workers receive only 61% of male full-time hourly wages, compared with a European average of 73%.
>
> The proposal is designed to produce greater pay transparency in businesses, which can be expected to resist the measure on grounds of cost and bureaucracy.
>
> The Commission will also call on the government to lead the way in reviewing pay differences, on the basis that many low-paid occupations dominated by women – including nursing, caring and cleaning jobs – are in the public sector.
>
> *The Guardian*, 27 February 2001

To answer this, a sort of 'reverse logic' to that of the confidence intervals in the previous section is used, and a check is made to see if the difference is outside the confidence intervals. If it is not, the difference may be due to the vagaries of sampling; it might be simply luck, and if you got another two samples a difference might not occur. If the difference between the two means is indeed outside the confidence intervals then this is unlikely to be due to chance, and something 'real' is happening.

The logic of the technique is similar to that applied to a single mean. Suppose that you took a large number of samples, each containing both male and female employees, and for each sample obtained a difference by subtracting the mean income for women from the men's mean income. These differences would have a pattern that followed a normal curve, just like the curve you looked at in the last section. You could call it the distribution of differences. And just as the mean of a large number of single sample means would be the mean of the population, so the mean of these differences would be the difference between the means of the two groups.

The hypothetical set of differences has both mean and standard deviation, just as the hypothetical set of means did in the last section, and the standard deviation of the hypothetical set of differences is also called the standard error.[7]

To calculate the standard error of the difference of the means, you calculate the standard error of each of the two means and then do arithmetic on them:

$$\text{standard error difference} = \sqrt{\text{first standard error}^2 + \text{second standard error}^2}$$

So, to get the standard error of the difference, square each of the two standard errors, sum them and calculate the square root of the total.

So with (estimated) standard deviations of £3.62 for men and £3.16 for women, then if the sample sizes are 600 and 400, respectively,[8] the two standard errors would be £0.40 and £0.39, respectively.[9] The standard error of the difference is then £0.56.[10] Now the logic is that the 95 per cent confidence level is two standard errors, which is £1.12 in this example. If the difference is more than this, then it is 95 per cent certain that it is not due to luck, and if the difference is more than two and a half times the standard error of the difference (£1.40), then it is 99 per cent certain that the difference is not due to chance. The difference between £9.82 and £7.88 is £1.94. So you can be 99 per cent certain that there is a real, substantive difference between the income of male and female employees in the population.

Difference between two percentages

The same technique can be used for comparing two proportions from a reasonably sized sample. Chapter 2 looked at an exit poll of 8,318 voters for the notorious US presidential elections in 2000, which gave Gore 48.4 per cent and Bush 48.8 per cent. The two standard errors are both 0.55 per cent.[11] The standard error of the difference is 0.78 per cent.[12] If the difference is more than two standard errors it is unlikely to have been due to chance. Two standard errors is 1.56 per cent, and since the difference is only 0.4 of a per cent, this could be due to chance rather than to a substantive difference between the two votes. The election was just too close to call.

However, the British general election seems to be another matter at the time of writing. The latest opinion poll of 1,900 voters gave the Conservative Party 30 per cent and Labour 50 per cent. The standard error for the Conservative figure is 1.05 per cent, and 1.14 per cent for Labour. This gives a standard error for the difference of 1.55 per cent, and $2\frac{1}{2}$ standard errors is 3.87 per cent. So if the difference is outside 3.87 per cent it is 99 per cent certain that it is not due to chance. The 20 per cent difference in the poll is way outside that, and most commentators view the outcome of the election as a foregone conclusion.

BOX 7 SUPPORT FOR CORPORAL PUNISHMENT

Parents call for schools to bring back the cane

A majority of parents want corporal punishment to be reintroduced in schools to tackle what they perceive is an increasing problem of classroom disorder, according to a poll published yesterday....

The survey of 1,000 parents in England and Wales... showed 51% of parents think reintroduction of corporal punishment is the answer to the problem. Corporal punishment was abolished

14 years ago throughout all state schools and in the private sector last year.

The Guardian, 8 January 2000

This report, which was carried by all the national press and television, shows well the importance of establishing reliability. The report states that 51 per cent of the sample of 1,000 was in favour of corporal punishment, which means that 49 per cent were against.

The standard errors of these percentages (which will be identical, since they sum to 100 per cent) are 1.58 per cent.[13] The standard error of the difference is therefore 2.24 per cent.[14]

The 95 per cent confidence interval is two standard errors, which is 4.48 per cent. If the difference is outside this interval, it is probably not due to chance. The actual difference is 2 per cent, which may well be due to chance, and if another 1,000 people were sampled the result may be different. The headline is misleading, and all the survey really tells us is that opinion is sharply divided.

One-tailed tests

The analysis of means, their differences and confidence intervals is just as receptive to a specification of direction as the measures of association were. If, for instance, in the previous example you are only considering whether men's income is higher than women's, the reliability of the confidence interval can be doubled. This has the effect of reducing the difference you can expect 99 per cent of the time to 2.33 standard errors, and 95 per cent of the differences would be within 1.66 standard errors, which in the example would be a difference of only £0.37 at the 1 per cent level and £0.27 at the 5 per cent level.

In fact, you can now see why it is called a one-tailed test. Only one side of the curve is considered, so only one of the 'tails' at the edge of the graph in which rogue samples fall is under consideration. The 5 or 1 per cent of the data that is in this tail will therefore be smaller than if both tails are considered.

Limitations

The techniques in this chapter are subject to two important limitations, which it is vital to bear in mind. First, they are designed to be used only for large samples and are inapplicable for samples of less than thirty. While most social surveys will deal with samples of several hundred or over a thousand, this limitation may be very relevant to the analysis of psychological experimental data, for example. The alternative tests to use are mentioned in Chapter 10, as well as modifications to measures such as the standard deviation.

Another major limitation, which applies to all samples, is that the analysis assumes that the sample is approximately normally distributed. If the sample distribution is clearly skewed, or if it is very flat or very peaked, the techniques in this chapter should not be used. In fact, none of the techniques for ratio-level data should be applied to such distributions, including Pearson's correlation and regression analysis. Use ordinal data techniques instead.

SUMMARY

- Questions of reliability are just as important for describing data as for investigating associations. You wish to know about a population rather than a sample.
- You cannot know about the population, but you can estimate the shape of the distribution of means from a large number of samples. This is called a *normal distribution*.
- Using this and the sample mean and standard deviation, you can estimate the boundaries within which you are confident the population mean will occur.
- You can use a similar technique on percentages.
- A 'reverse logic' is used to examine differences between two means or percentages. If the difference is outside of the confidence intervals, it is probably not due to chance.

APPENDIX

Standardised scores

Sometimes these scores are used to describe where a particular piece of data fits into a set.[15] It uses the properties of the normal curve to help to interpret any one particular score. The thing is that the area under the normal curve for any number of standard errors is known.[16] For example, while 95 per cent of the data will be within two standard errors, 68.26 per cent will be within one standard error: that is, 34.13 per cent within one standard error each side of the mean.

To take an example, one of the other measures of hospital performance published by *The Sunday Times* was the cardiology waiting list. This was the percentage of cardiology outpatients that had not been seen within thirteen weeks. The mean for this was 63 per cent, with a standard deviation of 24 per cent. One question it might be useful for individual trusts to answer is how high up the scale their particular indicator is. In fact, it is possible to use these two figures to interpret the position of any individual return. The method of interpreting this is to subtract the mean and divide by the standard deviation. This is called a *Z score*:

Z score = (number − mean)/standard deviation

Airedale Trust had a reported waiting list of 39 per cent. This is clearly below average, but it is not obvious how far down the scale its performance actually is. If its figure was 39 per cent, its Z score would be:

(39 − 63)/24

which is −1. Now, since 34 per cent of data in a normal distribution falls within about two standard deviations of the mean, 34 per cent of the data has a Z score of between −1 and +1. We also know that half the distribution is above and half below the mean, since that is what the mean measures. The Airedale figure is therefore in the bottom 16 per cent (50 − 34) of hospitals.

Bear in mind that these scores are valid only for normal distributions and do not apply to skewed data.

Notes

1. To be strictly accurate, this should be called the *standard error of the mean*.

2. Strictly speaking, the figures are 1.96 for 95 per cent and 2.58 for 99 per cent confidence, respectively.

3. The arithmetic is:

    $11.26/\sqrt{6,342} = 11.26/76.64 =$ **0.14**

4. The arithmetic is now:

    $SE = 11.26/\sqrt{50} = 11.26/7.07 =$ **1.59**

5. Strictly speaking, this is the *standard error of the proportion*.

6. $SE = \sqrt{33 \times (100 - 33)/7,830}$
    $= \sqrt{33 \times 67/7,830}$
    $= \sqrt{2,211/7,830}$
    $= \sqrt{0.28} =$ **0.53%**

7. Again, strictly speaking, this is the *standard error of the difference*.

8. The sample size is actually nearly a quarter of a million, which will give standard errors so small as to be negligible. The hypothetical samples used here are only for illustration.

9. The arithmetic is:

    $SE (men) = £9.82/\sqrt{600} = £0.40$

    $SE (women) = £7.88/\sqrt{400} = £0.39$

10. The arithmetic is:

    $SE \text{ difference} = \sqrt{0.4^2 + 0.15}$
    $= \sqrt{0.16 + 0.15} = \sqrt{0.31}$
    $= 0.56$

11. SE for Gore is $\sqrt{(51.6 \times 48.4)/8,318}$; for Bush it is $\sqrt{(51.2 \times 48.8)/8,318}$

12. 0.55^2 is 0.3. SE of difference is $\sqrt{(0.3 + 0.3)}$

13. $\sqrt{(51 \times 49)/1,000} = \sqrt{2.5} =$ **1.58**

14. $\sqrt{1.58^2 + 1.58^2} = \sqrt{2.5 + 2.5} = \sqrt{5} =$ **2.24**

15. They are also used in advanced regression analysis.

16. See Clegg, F., *Simple Statistics* (Cambridge: CUP, 1990) for a table.

Chapter 8
HOW TO PRESENT TABLES AND GRAPHS

While the focus of this book is on the communication of data, the emphasis so far has been on the analysis of data and on understanding the data presented by others in reports, newspapers and textbooks. This chapter will discuss how to communicate your analysis to others: an important and often overlooked aspect of data analysis. If your analysis cannot be communicated, it may as well not have been done.

It might surprise the reader when I suggest that data should *not* necessarily be presented in the same way as I have presented it in this book. The tables in the book were used for a different purpose; for exploration and instruction. The interest was in explaining the techniques, not in the data itself or the substantive issues it helped to shed light on. What this chapter will cover is the different reasons for presenting data and how to choose the best presentation technique for each function. It will then give some basic guidelines for the different types of data presentation.

While the communication of data is not difficult, it is useful to know a few basic principles, the first of which is that some effort is needed for the task. While not difficult, presenting data is not straightforward either. *Data does not speak for itself.* It always requires both analysis and interpretation, and it needs to be presented so that the reader can easily understand it.

Following from this, the first rule is: *always refer in the text to any data you present*. If you ignore your own data, then your readers are also likely to ignore it.[1] Furthermore, the text needs to indicate the story that the data tells. In fact, it may be a good idea to start with the text in order to think through the story that you think the data tells.

Data for explanation or storage?

Data can perform either of two functions:

1 *Reference function*: the reader retrieves the data and analyses it themselves; or

2 *Explanation function*: the data is used to support an argument or to show something up.

These functions are different, and the same table or graph should not be used to perform both tasks. If necessary, produce two tables and put the reference table in an appendix.

The two methods for presenting data are tables and graphs. Reference data is presented almost exclusively in table format, whereas explanation data may be either graphical or tabular. This chapter will concentrate more on the explanation function, although many of the principles for presenting tables also apply to reference data.

Graphs or tables?

While tables are more accurate, in that they can transmit exact magnitudes, graphs have more visual impact. Not only does this break up longer portions of text and make the piece more visually interesting, but it can also be easier to portray a story visually.

Graphs are good for:

- making text eye-catching
- making a point
- communicating to the general public
- showing trends and relationships.

Tables are good for:

- giving exact figures
- ensuring that the message is clearly and precisely understood
- ease of drafting and composition.

In general, tables are preferable unless there is a good reason for using a graph or chart.

GUIDELINES FOR TABLES

Table structure

All tables need to include:

- what is being measured and how it is categorised;
- units of measurement, coverage and time period; and
- source of data.

The source is often put at the foot of the table, the other information is usually in headings and subheadings. Table 51 is reproduced from Chapter 1. If there are different units in different parts of the table, label each part with the appropriate unit. In Table 52, from the research outlined in the next chapter, there are two variables, the level of grant and the proportion of exam passes in each local council. The per capita grant is in pounds, and the proportion is a percentage.

Table 51 Low incomes by region, England and Wales, 2000

Percentages

Region	Gross incomes below £200/wk
Wales	12.2%
N East	12
Yorks & Humberside	10.6
N West	10.5
E Mids	10.4
S West	10.3
W. Mids	9.1
Eastern	7.9
S East	7.4
London	5.1
All	9.2

Source: adapted from ONS website

Table 52 Council grant and exam performance, selected councils in England, 1997–99

Local council	Average RSG/cap 1996–99 £	GCSE average 1997–99 %
Barking and Dagenham	514	32.1
Barnet	325	55.1
Barnsley	337	30.3
Bath and North East Somerset	241	53.1
Bedfordshire	279	46.0
Bexley	339	46.7
Birmingham	544	36.4
Blackburn	522	35.9
Blackpool	376	35.6
Bolton	386	41.2
Bradford	483	30.2
Brent	594	42.0
Bristol	305	30.7
Bromley	258	54.9

Source: DTER and Education Dept websites

Tables are usually numbered in some sort of series, either simple progression from 1 upwards, or sometimes by chapter in longer publications (2.1, 2.2, and so on). This gives the table a tag by which it can be referred to in the text.

Two- and three-way tables

The heading for two- and three-way tables should show clearly both of the factors by which the data is being categorised, as was done in the tables in Chapter 2. Table 53 is reproduced from there.

Table 53 1997 income by 1991 income, UK adults

| 1991 Income group | 1998 income group | | |
	Top 40%	Lowest 60%	All
Top 40%	13 873	6 493	20 365
Lowest 60%	6 351	24 195	30 547
All	20 224	30 688	50 912

Source: estimated from *Social Trends* 30 (2000)

Table 54 British university applicants by occupation, gender and race, UK, 2000

White		Accepted			Rejected		
	Male	Female	Total	Male	Female	Total
Professional & intermediate	46%	54%	122 474	48%	52%	34 006
Skilled	44%	56%	57 390	46%	54%	22 905
Unskilled	44%	56%	19 009	47%	53%	8 625
All	*45%*	*55%*	*198 873*	*47%*	*53%*	*65 536*

Black		Accepted			Rejected		
	Male	Female	Total	Male	Female	Total
Professional & intermediate	42%	58%	3 294	43%	57%	1 796
Skilled	40%	60%	2 459	46%	54%	1 488
Unskilled	40%	60%	1 091	39%	61%	774
All	*41%*	*59%*	*6 844*	*43%*	*57%*	*4 058*

Asian		Accepted			Rejected		
	Male	Female	Total	Male	Female	Total
Professional & intermediate	51%	49%	9 659	54%	46%	3 113
Skilled	50%	50%	8 500	55%	45%	3 661
Unskilled	50%	50%	4 148	57%	43%	1 904
All	*51%*	*49%*	*22 307*	*55%*	*45%*	*8 678*

Source: adapted from UCAS website

Presenting three or more variables is difficult. If the table is simplified enough it may be possible to present a reasonable table for three variables, provided that the control variable does not have too many categories. As in Chapter 5, each of the categories of the control variable should be attached to a clearly marked section that reproduces the bivariate association, such as Table 54. The table is much more feasible if one or more of the variables can be dichotomised, so that the data can be simplified and reduced. The next section will discuss how to simplify data. If the table cannot be simplified, you could consider using a standardised table, as discussed in Chapter 5. Even if it is possible to present a table with a large number of variables, the amount of text needed to explain the trends that the data in the tables shows will be quite large.

League tables

The idea of league tables is fairly straightforward, and most are reasonably easy to understand. Much depends on the number of cases being presented.

Although scales, such as those used for attitude or satisfaction surveys, are ordinal variables they do not lend themselves to presentation in league table format. A simpler percentage table is easier to understand. The data from the questions in Figure 31 would be better presented in the format of Table 55 than in some sort of league table, although even Table 55 is too large and would benefit from further simplification.

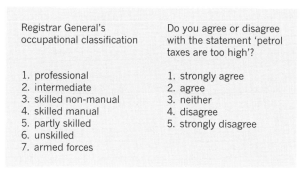

Registrar General's occupational classification	Do you agree or disagree with the statement 'petrol taxes are too high'?
1. professional 2. intermediate 3. skilled non-manual 4. skilled manual 5. partly skilled 6. unskilled 7. armed forces	1. strongly agree 2. agree 3. neither 4. disagree 5. strongly disagree

Figure 31 Two examples of ordinal data

Table 55 Taxation opinion by occupation, UK adults, 2001 (Percentages)
Do you agree or disagree with the statement 'petrol taxes are too high'?

Occupation	strongly agree	agree	neither	disagree	strongly disagree
Professional	5	25	17	23	30
Intermediate	7	24	21	25	23
Skilled non-manual	10	24	24	25	17
Skilled manual	12	25	24	23	16
Partly skilled	12	27	25	22	14
Unskilled	13	27	27	20	13
Armed forces	13	24	20	15	28
All	10	25	23	22	20

hypothetical data

For a true 'league table' of two variables where an index is used, or when a ratio variable is used as an index number, the problem will be how to select a small enough sub-sample of your data if the sample size is large. You could select this at random, or choose groups of, say, ten or fifteen cases. The data in Table 56 is from the full list of ninety-nine universities. Selecting the top, middle and bottom ten, and showing them side by side gives a smaller table that still manages to get the story of strong correlation across.

An obvious strategy for presenting these types of table is to sort the columns in order of one variable and present either the index/magnitude and/or the ranking (first to last) of the second variable. This enables comparisons to be made easily. This is also a good way of presenting ratio data, if a table is required instead of, or as well as, a graph. Indeed, the question of what order to present data is one the subjects of the next section on the presentation of data.

Table content

While the previous section was concerned with the overall structure of tables, this section looks at the body of the table. It will discuss how to simplify data so as to make the construction of tables simpler and will give some ideas as to the layout of tables to make the trends easy for the reader to pick out.

Layout and style

This is something of an art for explanation tables, and much depends on the purpose for which the table is to be used. Some basic principles:

- Put columns reasonably close together. If column headings are too long, shorten them or adjust the layout, rather than widen columns.
- Avoid vertical lines in the body of the table. Keep horizontal lines to a minimum.
- Print tables upright (portrait) on the page if at all possible.

Use typography for titles and subtitles (bold for titles and italics for subtitles seems an obvious scheme). For those of you using word

Table 56 Rankings and entrance requirements by university, UK, 2001

TOP 10 University name	Ranking	A-level points (100)	MID 10 University name	Ranking	A-level points (100)	BOTTOM 10 Institution	Ranking	A-level points (100)
University of Cambridge	1	98.8	University of Aberdeen	51	71.6	London Guildhall University	90	38.0
University of Oxford	2	97.6	University of Northumbria	52	53.2	University of Central England	91	45.2
London School of Economics	3	93.2	City University	53	70.8	University of Wolverhampton	92	39.6
Imperial College	4	92.8	Oxford Brookes University	54	50.8	Glasgow Caledonian University	93	50.0
University of York	5	83.2	University of Ulster	55	62.0	Thames Valley University	94	35.2
University College London	6	84.0	Bradford University	56	57.2	University of North London	95	29.2
University of St Andrews	7	77.6	University of Wales, Lampeter	57	41.6	South Bank University	96	39.2
University of Warwick	8	86.0	University of West of England	58	52.8	University of East London	97	39.6
University of Bath	9	84.4	Kingston University	59	45.2	Napier University	98	41.2
University of Nottingham	9	87.2	University of Plymouth	60	48.0	University of Paisley	99	42.4

Source: *The Times* University Tables

processors, titles are usually in a two-point (0.7 mm) larger font than the rest of the table. Eight- (2.8 mm) or maybe nine-point (3.2 mm) is a nice font size for tables, depending on the font used in the text. It is also common to use italics for percentages.

Data reduction and presentation

The point of explanation tables is to present the *results* of analysis, rather than just 'storing' the data. This means that the analysis is done by the author, not the reader, and that the table needs to be as simple as possible in order to be readily understood. However, there is often a trade-off between simplicity and accuracy, and just how simple the table should be made depends on the purpose and readership of the report. A report for a general readership, or a newspaper article, would trade off accuracy for simplicity to a much greater degree than an academic text or a student essay.

The following guidelines should be useful:

1 *Round off all numbers*: decide how much detail is needed in the table. Is it really necessary to know a percentage to two decimal places? If not, round off to one place or even no decimals, especially since whole numbers are easier to compare. Maybe frequencies could be rounded off to the nearest hundred or thousand. Be particularly harsh on spurious accuracy – there is little point in reporting analysis on measures with debatable accuracy or validity to three decimal points. If rounding off causes slight discrepancies in totalling, do not correct the discrepancy but report it as a note to the table. This often happens with percentages.

2 *Put numbers to be compared in columns*: it is easier to compare numbers vertically than horizontally. This usually means that the independent variable is put in the rows.

3 *If possible, arrange rows and columns in size order, preferably from highest to lowest*: this is part of the need to use the table to 'tell the story' of the data. By putting the data in order of dependent variable the trends become easier to spot. The exception to the above rule is where there is a 'natural' order for the independent categories, such as the mobility tables in Chapter 2. These would be tough to follow unless the data was presented in order. Also, most time-series data needs to be presented in chronological order.

4 *Include row and/or column aggregates and total sample size* (N): this has two purposes: (1) it enables the table to be reconstructed if percentages are used; (2) it acts as a reference to grasp the relative contribution of each figure. For the same reason, you could consider reporting averages, so that the data can be compared. If necessary, subtotals in a frequency table can be offset by a column, accountancy style. In Table 57, the subtotals are offset in a separate column. A subtotal within the subgroup is offset by two columns. Note also the use of italics, bold text and lines to aid clarity.

Table 57 Notifiable offences recorded by the police in England and Wales, by offence, April 1999 to March 2000

Offence category	Number of offences (thousands)	
Violence against the person	581	
Sexual offences	38	
Robbery	84	
Total violent crime		*703*
Burglary		
Burglary in a dwelling	443	
Burglary other than in a dwelling	464	
Total burglary		*906*
Theft & handling of stolen goods		
Theft from the person	76	
Theft of pedal cycle	131	
Theft from shops	292	
Theft from vehicle	669	
Theft of motor vehicle	375	
Total theft of & from vehicles		*1 044*
Vehicle interference and tampering	57	
Other theft & handling stolen goods	623	
Total theft & handling stolen goods		*2 224*
Fraud and forgery	335	
Criminal damage	946	
Drug offences	122	
Other notifiable offences	66	
TOTAL		*1 468*
Total all offences		**5 301**

Source: adapted from COI

Table 58 Direction of two-way tables – Voting by income, US exit poll, 2000

	Voting behaviour						Income	
Income	Gore	Bush	All		Voting behaviour	Below $20 000	Above $20 000	All
below $20 000	66%	34%	780	rather than	Gore	66%	48%	49.8%
above $20 000	48%	52%	7 308		Bush	34%	52%	50.2%
All	49.8%	50.2%	8 088		All	780	7 308	8 088

Source: adapted from *L.A. Times* exit poll

These rules usually mean that two-way tables should be percentaged across the rows rather than down the columns.

Table 58 compares a row percentage table with a column percentage. In the first table, the two percentages to be compared are then aligned vertically, which makes them easier to compare using mental arithmetic. Certainly, control variables are usually better stacked (like Table 54) instead of being presented side by side, since the values of epsilons (percentage difference) can then be compared. Stacking the tables also makes it easier to put the table in portrait orientation.

In addition to these techniques, the aim is to keep the table as small as possible. This involves simplifying and reducing the data. Simplifying the data involves altering it in some way. Calculating percentages is one obvious way, and comparing the frequencies with the mean is another. Looking at differences or changes in frequencies of percentages might be feasible, especially for time-series data. Time series also lend themselves well to index numbers, where each figure is divided by a base period and multiplied by 100 (and then, following rule 1 above, rounded off).

One current debate is the policy implications of an alleged large growth in the number of senior citizens. In Table 59, changing the frequencies to an index number based on 1961 gives a much clearer indication of the anticipated changes in the number of senior citizens in the UK. This technique is especially useful when comparing series of variables with different units or starting amounts to find out if one thing changes faster than another. Table 60 shows

Table 59 Constructing an index of senior citizens. 1991 onwards are projections

	Year	Senior Citizens (000s)		Year	Index	
Base Year →	1961	7 747		1961	100	= 100 × 9 123/7 747
	1971	9 123		1971	118	
	1981	10 031		1981	129	
	1991	10 597	becomes	1991	137	= 100 × 10 031/7 747
	2001	10 800		2001	139	
	2011	11 915		2011	154	
	2021	12 201		2021	157	

Source: adapted from *Annual Abstract of Statistics* HMSO (2000)

Table 60 Violent crime reported by police. Index: 1994 = 100

Year	England & Wales	Scotland
	303 742	23 382
1994	= 100	= 100
1995	105	104
1996	114	109
1997	116	104
1998	109	113

Source: adapted from *Annual Abstract of Statistics* HMSO (2000)

clearly that the violent crime rate was rising faster in Wales and England than in Scotland until 1998.

While there are a large variety of different ways in which data can be altered, it still needs to be readily understood by the reader. The more the data is transformed, the harder it is to recreate the original.

The aim is to replace the original data in the table with the new data. So (unlike in this book), if frequencies are percentaged off only the percentages would be put in, although (following rule 4) totals should be put in the row and column marginals, as with Table 58. If both frequencies and percentages must be included, either they should be put as separate tables or the two different units should be distinguished by typography, for example by putting percentages in italics, as in Table 61.

Table 61 1997 income by 1991 income, UK adults (with row percentages)

1991 income group	1997 income group				All
	Top 40%		Lowest 60%		
Top 40%	13 873	68%	6 493	32%	20 365 = *100%*
Lowest 60%	6 351	21%	24 195	79%	30 547 = *100%*
All	20 224	*40%*	30 688	*60%*	50 912

Source: estimated from *Social Trends* 30 (2000)

Table 62 1997 income by 1991 income, UK adults

1991 income group	1997 income group		All
	Top 40%		
Top 40%	13 873	68%	20 365
Lowest 60%	6 351	21%	30 547
All	20 224	*40%*	50 912

Source: estimated from *Social Trends* 30 (2000)

One way of reducing the size of tables is to show only one percentage of a dichotomous (two-choice) variable, as in Table 62, since the other percentage can be obtained by mental arithmetic. This can be useful in simplifying large tables.

The next step would be to deliberately make variables into dichotomies, by combining or omitting data in the manner described in Chapter 2. Do this with caution, however, since it destroys the data (consider putting a full reference table in an appendix). On the other hand, if dichotomising is the best or only way to communicate data, then so be it.

Reporting statistical significance

There is conflicting advice on this. This author feels that while reliability is an important part of data analysis, the significance is not especially part of the 'story' of the data. My inclination is to put the significance level as a footnote or possibly in italics outside the main body of the table. On the rare occasions where differing reliability levels need to be reported, a separate table in a footnote or appendix is probably justified, but few readers of this introductory textbook are likely to come across this.

GUIDELINES FOR CHARTS

Charts are good for giving visual impact to a piece of writing. They are also useful for encapsulating the main thrust of an argument in a way that numbers cannot. They also bring out relationships between variables quite well.

If charts are to communicate well, they need to be in a form that the reader will be familiar with and able to interpret. This means sticking to the basic forms of charts – pie and bar charts, and line and scattergraphs. There may also be a place for pictograms, especially if writing for a more general audience, but many of the elaborate features of computer packages are probably more distracting than useful.

Although the 3-D graph in Figure 32 is more eye-catching than the simple line graph in Figure 33, the latter is easier to follow and is more intuitive.

The labelling and titling of a chart is pretty much the same as for a table, as is the need to explain it in the text. Charts are numbered as a separate series from tables, so the first chart will still be 'Figure 1' even if it comes after 'Table 1'. Bear in mind that too much detail will defeat the object of a chart, which is to put across the data in a visual manner.

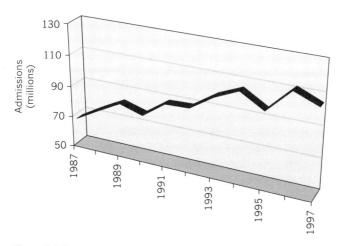

Figure 32 Cinema admissions: interest rather than simplicity

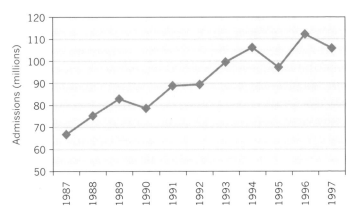

Figure 33 Cinema admissions: simplicity is usually best

General rules

- Make the chart the right size – not too small, but not too large either: a whole-page chart is too big.
- Avoid using label keys if possible – label straight onto the sections of the chart.
- If possible, include relevant numbers (percentages, frequencies).
- Do not put too much detail in charts.

Pie charts

These are useful for demonstrating proportions. Two or three pies can be put side by side to compare different variables, or as an alternative to a table, such as Figure 34, which is a highly visual alternative to Table 53. If using two or three pies, keep them the same size. Using more than three pies is likely to be counter-productive; it is difficult to make comparisons. Also, using many slices in the pie makes it tough to follow unless there is one dominant slice that you wish to draw to the reader's attention.

Usually, Figure 35 would be a poor pie chart, in that there are too many slices to even label properly. However, the point of the graph is the dominance of US usage. Another way of emphasising a particular proportion that you wish to draw attention to is to use an 'exploded' pie chart. Figure 36 highlights the high US usage.

HOW TO PRESENT TABLES AND GRAPHS 121

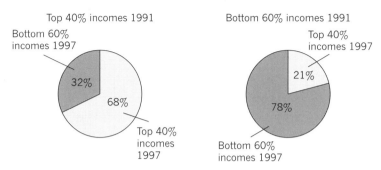

Figure 34 1997 income group by 1991 income group

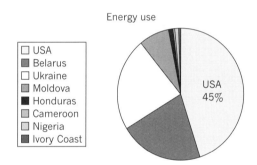

Figure 35 Energy use by selected countries

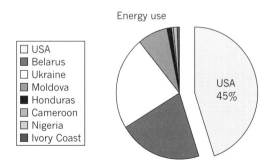

Figure 36 Energy use by selected countries

Bar charts

There are various types of bar chart, and something in the range can usually be found to cover most needs:

- *single bar charts* can be used to represent frequency tables;
- *grouped bar charts* can be used to compare several variables;
- *stacked bar charts* can be used to compare proportions;
- *line bar charts* can be used as an alternative to line graphs;
- *pictograms* can be used for extra visual impact;
- *histograms* show distributions.

The same general principles apply to these charts as to all others: correct size, differential shading, correct headings, units, coverage and direct labelling of categories. This last may necessitate horizontal rather than vertical bars if the labels are long.

Single bar charts

These are relatively straightforward. If the data is not ordinal or ratio, put the bars from highest to lowest (Figure 37), except for any catch-all 'other' category, which is put last. To compare two variables, use either two single bar charts side by side (Figure 38), or a grouped bar chart.

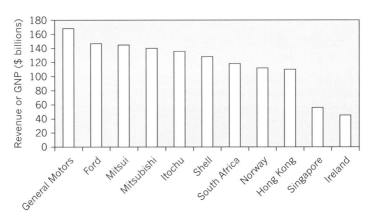

Figure 37 Income by selected organisations
Source: adapted from *World Guide 1999/2000*

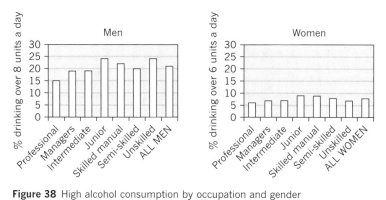

Figure 38 High alcohol consumption by occupation and gender
Source: adapted from *World Guide 1999/2000*

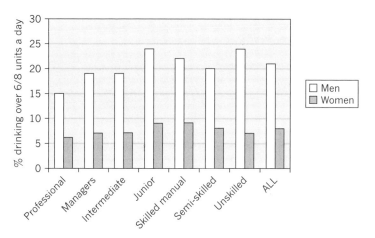

Figure 39 High alcohol consumption by occupation and gender
Source: adapted from *World Guide 1999/2000*

Grouped bar charts

These have groups of different shading for each category of the variable, as in Figure 39. To consider a control variable with only two or three categories, you can put two or three charts side by side as in Figure 38.

If there are several categories, resulting in many bars, make the bars thinner and the spaces between them bigger. If one bar is always smaller they could be overlapped, as in Figure 40.

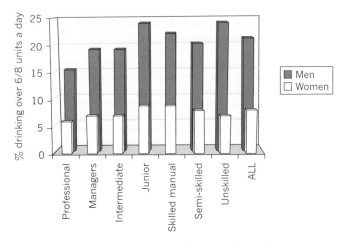

Figure 40 High alcohol consumption by occupation and gender
Source: adapted from *World Guide 1999/2000*

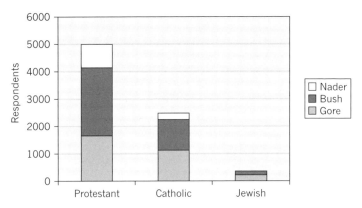

Figure 41 Voting behaviour by religion, US presidential election, 2000
Source: adapted from *L.A. Times* exit poll

Stacked bar charts

These are an alternative to pie charts for showing relative proportions; they show the 'subtotals' of the component parts of each bar. The point of Figure 41 is to show two elements – variations in the totals, and also variations in the subtotals. If the only thing of interest is to compare the proportions of the totals,[2] the graph can be

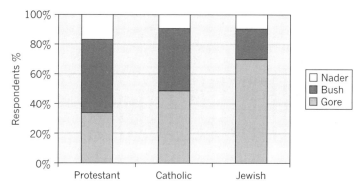

Figure 42 Voting behaviour by religion, US presidential election, 2000
Source: adapted from *L.A. Times* exit poll

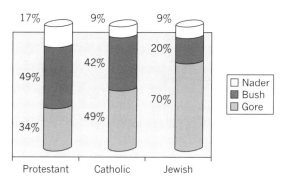

Figure 43 Voting behaviour by religion, US presidential election, 2000
Source: adapted from *L.A. Times* exit poll

redrawn as a percentage stacked bar chart, with all the bars the same height (100 per cent), as in Figure 42.

An eye-catching alternative is to use cylinder-shaped bars like those in Figure 43, or other typographical devices. As with all charts, the point is to figure out what message you wish to put across.

Line bar charts
These are useful as an alternative to line graphs, for example for showing trends (Figure 44). Finally, a variety of pictograms can add spice to your life and your text (see Figure 45).

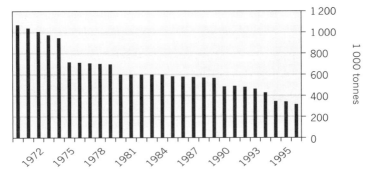

Figure 44 Trend of black smoke emissions, UK, 1970–97

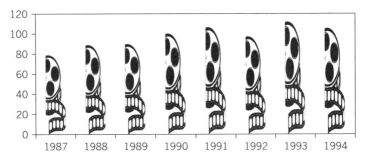

Figure 45 Cinema admissions

Histograms

These are bar charts where the bars are merged. They are used to show the distribution of data.[3] The histogram of Figure 46 shows a slight evening out of the skew on the right of the graph, but more from the third quartile to the first than evenly from right to left.

Scattergraphs and line graphs

Scattergraphs were used in Chapter 6 as a means of visualising correlation. The regression line may be added if the aim is to predict the dependent variable. It also acts as a reference to the graph. Figure 47, which is the same graph as in Figure 10 of Chapter 6, shows how

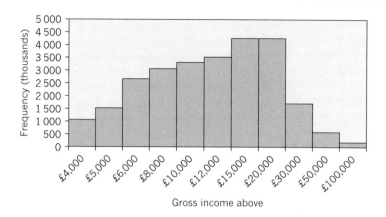

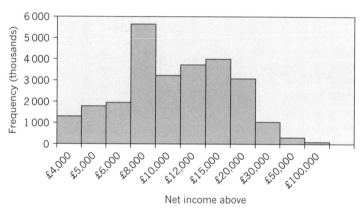

Figure 46 Distribution of gross and net incomes, UK taxpayers, 1997–98

adding the line of best fit helps the reader to see the fit of the data better. While this technique is especially useful for analysing time series, if prediction is not required the dots can be joined up in a line diagram, as with Figure 33.

Nonetheless, while all these pictures may tell a thousand words, you will still need some more words of your own to explain the story that you feel the graph tells; graphs no more speak for themselves than tables do.

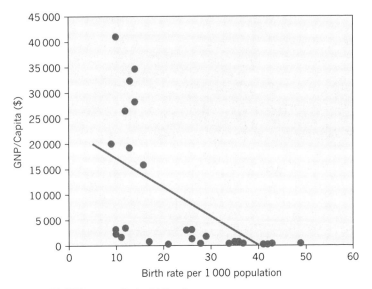

Figure 47 GNP per capita by birth rate
Source: adapted from *World Guide 1999/2000*

SUMMARY

- Always refer to graphs and tables in the text, and explain what they show.
- Tables are more exact; graphs and charts have more visual impact.

Tables

- Use tables for either explanation or reference. Do not try to use the same table for both purposes. Use an extra table in an appendix if necessary.
- Headings should show the objects being analysed, categories of analysis, units, coverage and data source.
- Put categories to be compared in columns.
- Arrange rows and columns in size order if possible.
- Include totals to enable original data to be recovered.

- Transform data to simplify tables. Keep tables as small as possible. Compromises between simplicity and accuracy are often needed. Show only one percentage of two-category variables. Consider merging or deleting categories in large tables.

Charts

- Use simple, well-known forms of graph.
- Choose the right type of chart for the 'story' you want to tell.
- Do not make the chart too large, and label directly onto components.
- Do not put too much detail in charts. It may be better to use two or three charts side by side to make comparisons.

Notes

1 This is a common (almost universal) error in publications at all levels. Apologies for the slight rant.

2 If the totals are the only item of interest, there would be no need for a stacked bar. A simple single bar chart would do.

3 If the groups are of unequal size, the *area* of the bar is supposed to be altered, although few graphics packages seem to be able to do this.

Chapter 9

PUTTING IT ALL TOGETHER

This book has now covered a range of simple yet powerful techniques that will enable the reader to both analyse and present data, and just as importantly to make sense of data presented by others. While there are a variety of more advanced techniques that can be used to bring meaning to data, the techniques outlined in this book should be sufficient for a considerable slice of the data analysis problems that most social scientists are likely to come across. A brief outline of some of the more advanced techniques is given in the next chapter. This chapter attempts to place data analysis in the context of a larger research project and show how it fits into the wider research arena. Data analysis is not an end in itself but is simply one of many tools than can be used to help the understanding of the world around us.

The first part of the chapter gives an outline of an ongoing research project into the secondary school performance tables in England. This project is ideal as an illustration of introductory data analysis because it arose from an undergraduate course on research methods that I taught a few years ago. The project was initially simply an exercise for my students, but the results were so interesting that I developed it into a full-blown project.

Unfortunately, the research does not involve an element vital to many projects, in that no sampling is involved, so the second part of this chapter covers the basics of sampling techniques and their place in data analysis.

Overview: points to note

When looking through the case study, the following points might be of particular interest:

1 *Number crunching is a relatively small part of the report.* Even though the data analysis is a vital component of the research and of the argument, it is only a minor component of the final report. Although the statistical techniques may seem daunting when you first encounter them, familiarity with them tends to diminish their interest. The main part of the data analysis is interpreting what the results tell us about the question at hand.

2 *Much of the report concerns the background information, and the main element is the literature review.* The review of relevant publications gives a theoretical framework without which the data analysis is insufficient. Without theory, the researcher does not know where to look, what to look for or how to interpret results.

3 *There is as much interest in the validity of the measures as in the reliability.* While statistical techniques are concerned with reliable data, the study is also concerned with valid data. Reliability is a question of the replicability of the study, whereas validity is concerned with whether the data actually measures what it purports to measure. Arriving at a valid measurement of the area under examination is a major part of the study.

CASE STUDY
CASH FOR ANSWERS: AN INVESTIGATION INTO SCHOOL PERFORMANCE TABLES[1]

Background

The publication of secondary school league tables was part of a critical review of all public services that preoccupied the Thatcher and Major administrations of the 1980s and 1990s. It was first suggested in the mid-1980s and included in the 1988 Education Act. The first results came out in 1992.

The introduction of these performance tables was part of a move towards a 'quasi-market' approach in all areas of public service, including health and housing as well as education.[2] 'Quasi-market' mechanisms attempt to introduce elements of markets, such as competition and consumer choice, into public services while retaining overall government regulation.[3] The aim appeared to be to provide financial incentives to schools to take more notice of the wishes of parents. This was to be achieved by:

1 publishing exam results;

2 increasing choice for parents; and

3 giving schools direct control over budgets.

The idea was that schools that had poor exam results would lose pupils and income to those that did better, and they would therefore change their workplace practices to improve their results.[4]

The exam performance tables are a vital part of this scheme, because these are supposed to give parents the necessary information to make choices between schools. However, they are only going to be useful to parents in this way if they are not influenced by external factors outside of schools. The whole basis of the scheme was that the decisions of schools and local education authorities (LEAs), and the performance of educational professionals, was the causal factor of the exam results.

It was this assumption that the study investigated. The research considered the possibility that the exam results reflected the socio-economic background of the pupils, rather than the efforts of the teachers and other professionals. This would mean that the quasi-market model of parental choice would fail, since all the exam results told us was the background of the parents who were doing the choosing.

This was a criticism that had been voiced since the first publication of the tables, and many studies found high correlations between the results of individual schools and the background of their parents.[5] The present study looked not at individual schools but at local education authorities.

Research Method

The idea was to use payments to local councils as a measure of socio-economic background and compare this with the performance measures of the equivalent LEA.

Local councils are elected bodies charged with running local services such as refuse collection, road maintenance and education, the latter being run through LEAs. Councils are funded partly by local taxation and partly by a grant (revenue support grant: RSG) from the national government. These grants ranged between £25 million and £582 million in the year April 1998 to April 1999, with a median of £103 million.

The RSG is designed to cover the projected shortfall in income between the expenditure it is judged local councils will need to run public services and the potential income from local taxes. It is therefore a crude measure of local need and of socio-economic status.[6] The Education Department publishes five measures of educational performance: five high GCSE exam passes; five passes at any grade; an exam 'points' measure; no passes; and truancy rates. If these education measures show a higher degree of correlation with the RSG, which is being used as a measure of need, this would suggest that need has a major influence on exam performance.

Analysis

This summary will be confined to the highest measure of educational performance published by the Department: the proportion of pupils passing five GCSE exams at grade C or above. This would enable the pupils to take A-level exams, a prerequisite for university. The proportions range from 61 per cent for the schools in the highest council to 23 per cent in the lowest, with a midpoint of 45 per cent.

It was decided to analyse the data at the ordinal rather than the ratio level, even though it is ratio-level data. There were several reasons for this. First, the interest is in the idea of 'league tables', of the relative ranking of councils and LEAs, rather than whether the magnitudes themselves matched. Second, the data – particularly the

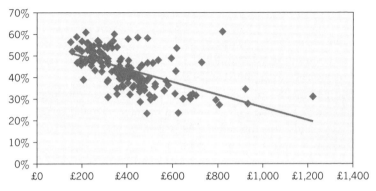

Figure 48 High exam passes by RSG, 1999

Table 63 Five 'high' GCSE passes by RSG per capita, England, 1996–99

	1996–97	1997–98	1998–99
Rank correlation r_s	−70.9%	−71.9%	−64.8%
No. LEAs	107	128	147

All results significant at the 1% level
Source: adapted from DETR and Education Department websites

RSG – was subject to many outside influences. Finally, the idea of a league table is far easier to communicate.

If this is plotted on a scattergraph, the correlation can be clearly seen. Figure 48 shows the level of RSG per capita for local councils compared with the proportion of five high passes for the corresponding LEA. The closer to the line the higher the correlation. In fact, as Table 63 shows, the correlations over the last three years range from 65 to 72 per cent.

Comparing the pass rate in 1999 with the RSG gives a rank correlation of just under −65 per cent. That is to say, a high proportion of the high examination pass rate can be explained by the relationship between socio-economic status and education. The result is 99 per cent reliable.

The correlation is a negative – the more money the council gets, the lower the proportion of exam passes. This seems contradictory, but when it it is remembered that the grant is given to compensate for a shortfall in income the negative association makes more sense, since it means the higher the shortfall the lower the results, which might be expected.

Table 64 Frequencies of significant changes in LEA pass rates, England, 1998–99

Size of change 1998–99	5 high passes (A–C)	No passes
0–2%	21	56
2–4%	16	14
5–7%	6	2
Total	43	72
No. of LEAs	120	147

Recently the Education Department has tried to promote alternative measures, in particular the annual change in pass rates (which it calls the 'improvement measure'). One way to analyse these is to look at the difference between two percentages, using time as a variable. We could expect a variation of two standard errors in the usual course of events and would only regard the difference as substantial if it was higher than this. As Table 64 shows, only about a third of the changes in pass rates in the last year were significant (43 out of 120 LEAs), and of these only six were above 5 per cent. This measure therefore tells us very little.

In conclusion, the education league tables, which were designed to inform parental choice about school performance, and thereby to motivate poorly performing schools to increase their efforts, can be shown to reflect mainly the backgrounds of its pupils, rather than anything occurring within schools themselves. Furthermore, the alternative 'improvement measure', based on annual changes in the pass rates, is not a reliable measure.

SAMPLING TECHNIQUES

- **Title** Sampling techniques
- **What are they?** Methods for selecting a smaller group of people from a larger target population
- **Why use them?** 1 The reliability measures assume a random sample

    2 Need to know sampling methods because the practicalities of 'real life' sampling may necessitate weightings for data

Unfortunately, the study of school league tables does not include sampling, which is an important topic even if the data analysis is from data already collected (secondary analysis). This is because the reliability tests assume a random sample. If a non-random sample is used, reliability is compromised.

Random and non-random sampling

Sampling techniques can be divided into two types, probability and non-probability sampling. The former involves some form of random selection; the latter involves some form of selection of respondents by the researchers.

Probability sampling

This is the form of sampling that is usually associated with survey research, and with the quantitative, numerical data analysis that is the subject of this book. The idea behind a probability sample is that everyone in the study population has an equal chance of being sampled.

The process of probability sampling involves the following elements:

- defining a population;
- obtaining a sample frame; and
- choosing a sample from the sample frame.

Population

This is the group that the research aims to find out about. The population needs to be clearly defined: who exactly is the subject of the investigation? Is it everyone in the country? Or in an area? Or only adults? Or only a particular group?

The definition of the population that is to be researched will interact with the sampling process, since it determines who is to be included in the process. All members of the study population must be included in the selection process. By the same token, the people who are actually included in the group from which the sample is drawn determine who the conclusions of the research actually apply to. If a group is excluded – accidentally or on purpose – from the sampling process, then the results of the research do not apply to them. You need to ensure that all the people you wish to look at (and only these people) are included in the sampling procedure.

Sample frame

Having defined a population for study, probability sampling techniques then involve various methods of selecting at random a sample from this population. However, this choice invariably involves choosing at random from some kind of list of the population. This list is called a sampling frame.

Problems frequently occur with this frame. Often the populations that are under study are large, and the frame can be unwieldy to handle, even with the techniques that will be discussed later. A national UK frame would have 56 million names, maybe half of them adults. A sampling frame of all the voters in one parliamentary constituency might contain 70,000 names, with possibly 5,000 names in each local ward. Simply compiling such lists, or even using an already compiled list, is far from straightforward.

In addition, lists of this size inevitably contain inaccuracies, especially as they cannot be constantly updated. Furthermore, using lists that are already in existence can be a problem, since the list was created for other purposes and may not reflect the population that the study wishes to consider. The telephone directory, for example, contains most telephone users, whereas the electoral register contains the names and addresses of eligible voters. These populations are not the same.

Choosing a sample

In order to help with these problems, and in order to cut back on the resources needed to actually survey the sample once chosen, a variety of different survey techniques have been developed (see Table 65):

- *Simple random sampling* is, as the name implies, a straightforward random selection from a single list of the entire population. As such, it is easy to understand but, as was suggested above, hard to actually implement. The physical process of random selection can

Table 65 Types of probability sampling

Type	Frame	Method
Simple random	whole population	completely random – 'lottery' fashion
Systematic random	whole population	regular intervals + random start point
Stratified sample	one/several random subgroups	sample from subgroups
Cluster sample	list of natural units in population	all members of randomly chosen unit sampled
Multi-stage	combination of above in succession	

even be difficult. For this reason, even if a sample frame of the entire population is used, the sample is usually chosen systematically – for example, every 500th or 1,000th name is chosen.[7] This last is called *systematic random sampling*. Note that if systematic random sampling is employed, the starting point must be chosen at random, rather than always at the start of the list.

- *Stratified sampling* involves dividing the sample frame into groups according to relevant criteria, such as age, income, and so on. These groups are called strata. The sample is then drawn from all the strata. This ensures representation from all relevant groups in the target population.

One area of current debate, for example, is that black people in the UK are proportionately more likely to be the victims of crime than white residents. However, if a survey of views on crime used a simple random sample, the resultant example would be likely to have fewer black respondents, since there are fewer in the population. If this number is further divided by gender and/or age, the resultant sample may be so small as to be statistically unreliable. Stratification ensures that all relevant groups have a sufficient representation to give results that are reliable.

- *Cluster sampling* is similar to stratified sampling. Here the target population is grouped into 'natural' units, such as a geographical area. A number of these units are chosen at random, and everyone in the chosen units is included in the sample.

This is particularly useful for concentrating the survey in a few small geographical areas. For example, a US survey covering the entire country would be extremely resource-intensive to conduct. Apart from a massive sampling frame, either interviewers would have to travel huge distances or a large number of interviewers would have to be recruited, trained and briefed. Such a survey would realistically be limited to either postal surveys (which have a lower response rate) or telephone surveys (which are less accurate).

An alternative procedure is to pick a few smaller geographical areas at random and base the survey only on these areas. The process is still random, as anyone in the target population can be included, but the survey is much easier to conduct since only a small number of areas need be visited. Survey samples in the UK are often based on postcodes, which contain a small number of adjacent premises.

Another advantage of cluster sampling is that it is unnecessary to compile a complete sampling frame of the entire target population. Only a list of the clusters need be drawn up (a list of postcodes for example), and the sample list can be compiled after the clusters are picked.

- *Multi-stage sampling* uses a combination of these techniques in succession. So, for example, clusters may be chosen initially, and then either stratification or systematic sampling – or even a combination of both – might be used. There may also be several 'layers' of clustering.

Weighting factors

While the use of stratification and clustering in sampling may simplify the conducting of a survey, they do introduce a complication for data analysis. The entire reason for using probability sampling was that everyone should have an equal chance of being included in the sample. However, if group-based techniques such as clustering or stratification are employed, this is no longer the case (indeed, this is the whole point of stratification, as was explained above).

For example, if clustering is used, it is quite possible that the clusters will be different in size. If all the clusters have an equal chance of selection, then those respondents in the larger clusters have a lower chance of selection. The same would be true of strata.

This can be corrected for either when selecting the sample, or afterwards by adding a weighting factor to the data, to 'add more weight' to the unequal data. This factor is calculated by dividing the proportion of the group (cluster or strata) in the population by the proportion in the sample:

Weight = population % / sample %

As an example, assume a study on policing is being conducted. A stratified sample by race and gender is used to ensure that enough Indian, Pakistani and Bangladeshi respondents are polled for the results to be reliable, and 15 per cent of the sample respondents are so designated. However, in the last census only 3 per cent of the population was from these ethnic groups. In the data analysis, each of the relevant respondents should have their data multiplied by

3 / 15 = 0.2

in order to correct the over-representation caused by stratification.

Non-probability Sampling

While most survey sampling uses probability sampling, there may be several reasons why non-probability sampling is used:

- where resources – especially time – are inadequate for a probability sample;
- where no sampling frame exists;
- where access to respondents is difficult;
- at the beginning of a research project, when the existence of a research question or issue is being explored;
- where the research interest is in the subjects themselves, rather than in the general population.

Where these constraints exist, it is often considered that qualitative methods such as in-depth interviews or observation are more likely to be productive. Conducting a survey, for instance, on the homeless (no survey frame) or criminal gangs (no frame or easy access) is unlikely to yield reliable and useful results. However, surveys do sometimes use non-probability samples to cut time and costs: a quota sample, for example, can be done quickly and is popular with opinion pollsters.

Types of non-probability sampling

- Quota   cases selected by proportion of strata
- Convenience   cases selected by availability
- Similar/dissimilar cases selected by similarity or dissimilarity
- Typical cases   cases selected as typical examples
- Critical cases   cases pre-judged to be generalisable
- Snowball   selected interviewees suggest further people for sampling.

In survey work, only convenience and quota sampling are commonly used. A quota sample is similar to a stratification sample. A number of characteristics that it is believed are significant in the population are identified, and the population proportions displaying each characteristic are obtained. This proportion of respondents is then sought to fill the sample. So, for example, if 3 per cent of the population is Indian, Pakistani or Bangladeshi, then 3 per cent of the sample is the same. If the sample is 1,000, this gives a quota of thirty to be surveyed.

Another common form of non-probability sampling is convenience sampling: selecting respondents by availability. This often takes the form of 'intercept' sampling, whereby respondents are stopped in public places and asked to complete a survey. While many of the opinion pollsters defend quota sampling, the method does effect the reliability of results. Its use is therefore a matter for judgement. As a rule of thumb, if a non-probability sample is used, use only two-thirds of the sample size when calculating statistical reliability.

The other sampling techniques are far more common in qualitative analysis, where the emphasis is on validity rather than reliability. Focusing on cases that are atypical or extreme is not uncommon in this approach, especially if the interest is in the cases themselves rather than the population (why are these people different, and what can the differences tell us?). Snowball sampling may be the only way of obtaining access to respondents in some projects. Not all questions are suitable for analysis by numbers.

SUMMARY

- Reliability measures assume a random sample in which everybody in the study population has an equal chance of participating.
- Sampling involves defining the population, obtaining a sample frame and then choosing the sample from it.
- It is often more convenient to divide the population into groups and then choose at random between or within the groups, especially for large, widely spread populations.
- Quota samples are sometimes used for surveys, especially opinion polls. However, they are generally regarded as less reliable.

Notes

1 The full paper was published in *Sociological Research Online*, November 2000.

2 Levacic, R. and Hardman, J., Competing for resources: the impact of social disadvantage and other factors on English secondary schools' financial performance, *Oxford Review of Education*, vol. 24, no. 3 (1988) pp. 303–28.

3 LeGrand, J. and Bartlett, W. (eds) *Quasi-Markets and Social Policy* (London: Macmillan, 1993).

4 Levacic, R. and Hardman, J., *op. cit.*, p. 304; Bradley, S. and Taylor, J., The effect of school size on exam performance in secondary schools, *Oxford Bulletin of Economics and Statistics*, vol. 60, no. 3 (1998) pp. 291–324; Bartlett, W., Quasi markets and educational reforms, in LeGrand and Bartlett (eds) *Quasi Markets and Social Policy* (London: Macmillan, 1993).

5 e.g. McCullum, I. and Tuxford, G., Counting the context, *Education*, 26 November 1993, p. 400.

6 The full study uses two other measures based on council finance as checks.

7 The actual interval is the size of the sample frame divided by the chosen sample size. So to get a sample of 1,500 from a frame of $1\frac{1}{2}$ million entries, you would choose every 1,000th entry.

Chapter 10

SUMMARY

Table 66 Summary of statistical tests

Level of measurement	Description (one variable)		Association (more than one variable)	
	Measure	Reliability	Measure	Reliability
Nominal	1. *Central tendency*; mode 2. *Dispersion*: Report %	–	1. % difference (epsilon) 2. Cramers' *V*	chi-square
Ordinal	1. *Central tendency*: median 2. *Dispersion*: • Range • Inter-quartile range	–	Spearman's rho	(look-up table)
Ratio	1. *Central tendency*: mean 2. *Dispersion*: standard deviation; variation coefficient	*Z* test	Pearson's *r*	(look-up table)

The straightforward techniques outlined in this book should cover a considerable amount of the data analysis that both social scientists and professionals will require in analysing and communicating data. Just as importantly, a familiarity with these techniques will aid an understanding of much of the data that you will come across in other publications. You may well find yourself 'talking back' to data instead of just turning the page when you come across a table of figures.

Later in this chapter some of the more advanced techniques, which build on those in this book, will be outlined. These are meant for those readers who wish to take this study of statistical analysis further, which does not in any sense imply that the basic techniques covered above are inadequate. However, in this introductory book I have taken a few liberties in simplifying some of the formulas and techniques in order to make them easier to understand. Before I go on to discuss more advanced data analysis, I would like briefly to recap the techniques that have been outlined and explain the extra precautions that you may need to take. These revolve around adjustments to be made for small samples. While most social scientists will not need to take much notice of these, since they typically use samples of several hundreds or thousands, analysis of areas such as psychology-based experiments, medical statistics or quality control will frequently use small samples, of perhaps less than a dozen. These small samples may require adjustments to the formulas or changes to the techniques of analysis.

Summary of statistical tests and adjustments for small samples

Small-sample standard deviation

We began our odyssey into the world of data analysis by seeing how we could describe a group of data – how we could describe many numbers with one number. This involved measuring the centre of the group: the mode, median or mean. These are the three types of 'average' in common usage: the 'typical' case, the middle case and the arithmetical average of cases. However, we also need to describe how good a descriptor the 'average' is; we need a measure of spread. For the median this was the inter-quartile range; for the mean this was the standard deviation and the coefficient of variance.

However, the calculation of the standard deviation for a small sample is slightly different. The aggregate variation from the mean is divided by $N - 1$, not N. This measure is given a different Greek letter – sigma (σ):

$$\sigma = \sqrt{\text{sum(differences from mean)}^2 / (n - 1)}$$

This makes no odds to a sample of 1,000, but it does matter for smaller samples.

We then went on to look at the measurement of association between two variables, called bivariate analysis. For nominal-level data, the technique is to draw up a two-way table (sometimes called *cross-tabulating*) and to use a percentage difference (epsilon) on 2 × 2 tables and Cramer's *V* on larger tables or when exactness is required. For ordinal data, Spearman's rho was recommended as a measure, and for ratio data Pearson's *r* was suggested. It is also possible to predict one ratio level variable from another using regression analysis, which is particularly useful for analysing trends over time.

Small-sample reliability

The difficulties occur when checking for reliability. The technique of confidence intervals is used for considering the reliability of means and the difference between two means. However, as was mentioned in Chapter 7, the technique outlined was based on a normal distribution, which is not appropriate for samples of fewer than thirty. For these, a different distribution, called the t distribution, is used, and the technique is called a t test. The t distribution is slightly flatter than the normal distribution, and the confidence intervals slightly wider, but unlike the normal distribution the t curve changes with sample size. With larger sample sizes (over thirty), the t distribution is like the normal distribution. The t test is therefore a bit more complex than using the normal curve, so it was omitted from this book.

The tests for reliability of association at the ratio and ordinal levels of measurement were fairly straightforward. However, the chi-square test for category data is also vulnerable to small sample size, and if the sample is less than twenty, *Fisher's exact test* is generally used instead. Also, in small samples *Yates's correction* is sometimes used, whereby 0.5 is subtracted from the expected frequencies. Of course, for larger samples this will make little difference.

Gaining more knowledge

As can be seen from Table 66 categorising the techniques displayed in this book, the idea was to show one technique for measuring association for each level of data. Many more techniques are available, especially for nominal and ordinal association. Also, there are other measures for inter-level association, whereby data at one level is compared with data at another. The reasoning behind these particular choices is that they are simple, and the results all have the same format – a percentage indicating the strength of association. This is not the case with some of the alternative measures, but they do have other strengths. A list of further reading is included at the end of this chapter, and you can explore other statistical measures there.

Some of the other techniques that were mentioned earlier, or that develop the techniques in this book, might also be of interest.

Path analysis/causal modelling

This technique develops the multivariate analysis of Chapter 5. Essentially, it attempts to use the results of the multivariate tables to try to create a 'flow diagram', which shows how the dependent variables are *caused* and how much they contribute to the overall effect. It therefore goes beyond straightforward analysis and merges data with theory and experience.

Causal path models can be complex, since the number, directions and magnitude of variables can have many permutations.

Multiple regression

This book covered multivariate analysis only at the nominal level. However, regression techniques are also expandable for a larger number of variables, which is called multiple regression. This enables the prediction of a dependent variable value given the values of several independent variables, so it is a powerful technique.

Log-linear modelling enables the examination of contingency tables with a large number of variables.

LIMITATIONS OF NUMBERS: A WORD OF WARNING
READ THIS BIT

It might seem strange ending this book on a note of discouragement rather than encouragement, but I would feel guilty launching hordes of readers out into the world of data analysis without warning of some of the pitfalls. The limitations of quantitative data analysis are many in number and high in importance. Just as it is not the case that you need a degree in maths to analyse data, so it is also true that analysing data will not get you a degree in social science, psychology or anything else by itself. Neither by itself will data analysis yield much in the way of answers. Data analysis is simply one tool among many, and it is only useful under certain circumstances, circumstances that I would suggest are actually quite limited.

In particular, statistics will not

- tell us what causes something to happen;
- give answers to 'why' questions;
- tell us what questions to ask;
- stop us making fools of ourselves.

This last happens often. If you ask stupid questions you will get stupid answers, and if the associations you are searching for do not have a good grounding in theoretical analysis, or are not valid measurements of the things you are trying to measure, then you will end up talking nonsense. It may be reliable nonsense, but it will be nonsense nonetheless. Data analysis is too often used as an excuse to stop thinking.

However, when handled well, data analysis can be useful to debunk nonsense and can help us to find out about the world we live in. I hope this book has shown how that can be done and has given the reader knowledge, confidence and enthusiasm to take part in these explorations.

FURTHER READING

In the introduction I stated that the aim of this book was to bring the reader to a level where most introductory textbooks start. There are actually a great many textbooks on introductory data analysis, so just go into a good bookshop and pick one you like the look of. Of particular note if you wish to use SPSS is the series on data analysis published by SPSS. I have *The SPSS Guide to Data Analysis for*

SPSSx, Marija J. Norusis (Chicago: SPSS Inc., 1998), but this is now out of print. A new edition is issued with every new version of SPSS, and you should get the latest edition. I'm not a fan of SPSS (which is why I've set up *www.figuringfigures.com*), but because the book is linked to software it focuses on interpretation rather than calculation.

If you use SPSS in part of your course, you might also look at P.R. Kinnear and C.D. Gray *SPSS for Windows Made Simple* (Hove: Lawrence Erlbaum, 1994).

Another text of interest is Francis Clegg, *Simple Statistics* (Cambridge: CUP, 1990), which has useful appendices about statistical calculation.

Chapter 2
A useful text is C. Marsh, *Exploring Data* (Cambridge: Polity Press, 1988), chapter 7. Norusis also outlines some of the alternative measures of association for category variables.

Chapter 3
Clegg chapter 9 considers alternative tests for ordinal data, as does P.R. Hinton, *Statistics Explained* (London: Routledge, 1995), chapters 16 and 17.

Chapter 5
There are other techniques for standardisation. For an example of the use of standardisation in research see Marsh.

Chapter 7
For a good introductory text you could do worse than D. Rowntree, *Statistics Without Tears* (Harmondsworth: Penguin, 1981), chapters 4–7. Norusis is also good on this.

Chapter 8
This chapter leant heavily on Chapman and Mahon, *Plain Figures* (London: HMSO, 1997). This is an excellent book, but far too expensive for most people. One to borrow from the library.

Chapter 9

A text that tries to link surveys and data analysis is D.A. DeVauss, *Surveys in Social Research* (Englewood Cliffs, NJ: Prentice Hall, 1991).

Chapter 10

Path analysis

A good booklet is James A. Davis, *The Logic of Causal Order* (London: Sage, 1985), number 55 in the Quantitative Applications in the Social Sciences series.

Multiple regression

Two other Sage pamphlets are quite good: Schroeder, Sjoquist and Stephan, *Understanding Regression Analysis* (London: Sage, 1986), Quantitative Applications in the Social Sciences no. 57; and William D. Berry, *Understanding Regression Assumptions* (London: Sage, 1993), Quantitative Applications in the Social Sciences no. 92.

The section in Hinton is also useful, as is chapter 10 of A. Bryman and D. Cramer *Quantitative Data Analysis with SPSS for Windows* (London: Routledge, 1997).

Log-linear analysis

Try G.W. Gilbert, *Analysing Tabular Data* (London: UCL, 1993).

This list is by no means exhaustive, and it really is a case of checking out a range of texts for yourself.

Index

NB. Footnotes are designated by the letter n after the page number.

advanced techniques 143, 144–5
aggregates, columns/rows 115
agreement/disagreement responses 30, 34
analysis
  bivariate 25
  case studies 133–5
  de-mystification 2
  limitations 146–7
  multivariate 25, 60–9
  path analysis 66, 68, 149
  quantitative data 147
  regression analysis 86–90
applications, university places 66–9
arrangements, columns/rows 114–15
association
  between variables 24
  category variables 57–71
  causality 25–31
  Cramer V 57, 58–60, 70–1
  direction 37, 72
  independence 25–31
  league tables 33–43
  measures 24–32, 45
  more than two things 57–71
  multivariate analysis 60–9
  reliability confusion 45
  reliability differences 58
  strength 72
  see also category association

averages
  moving 89
  term usage 7, 12
  see also means
axis intercepts 83–5

background information, reports 131
bar charts 122–6
  grouped 123–4
  single 122–3
  stacked 124–5
bivariate analysis 25
bivariate tables 63–4
bold text, use of 115

calculation
  confidence intervals 97
  Cramer V 70–1
  linear regression 92n
  $r(rho)_s$ correlation 41–2
  standard error difference 100
'caning in schools' article 101–2
'care for elderly' article 87
case studies
  'Cash for Answers' 131–5
  hospital performance 40
  overview 131
  performance tables 40
  points to note 131
  rank correlation 40
  school performance tables 130–5

INDEX

'Cash for Answers' case study 131–5
  analysis 133–5
  research method 133
category association 57–60
category data reliability 46–9
category tables 48–9
category variables 57–71
causal modelling 146
causal paths 70n
causality association 25–31
cells, tables 26–7
central limit theorem 95
central tendency, measures 5, 6–19
charts
  bar charts 122–6
  communication 3
  construction guidelines 119–28
  general rules 120
  line bar charts 125–6
  pie charts 120–1
  stacked bars 124–5
chi-square
  calculation 52–6
  degrees of freedom 53
  expected frequencies 52
  minimum values 54
  reliability 47, 54
  using 54–5
cluster sampling 138–9
coefficient of variance 18–19
columns
  aggregates 115
  arrangements 114–15
  marginals 117
  used for comparisons 114
comparisons, columns use 114
confidence intervals 93, 96–8
  calculation 97
  portions 98
  'reverse logic' 99–100
contingency tables 32
control variables 61–2
  examples 63
convenience sampling 141

correlation
  definition 33
  graphs 73
  logic 73
  negative 76, 134
  perfect 76
  positive 76
  product moment 76–7, 79
  rank correlation 33–43
  zero 76, 77
Cramer V
  association 57, 58–60
  calculation 70–1
  disadvantages 59
cross-classification 28
cross-tabulations 28, 144
cumulative percentages 12

data
  analysis de-mystification 2
  functions 107
  presentation 106, 114–18
  reduction 114–18
  as reference 107
  spread restrictions 80–1
  storage 107
de-mystification, data analysis 2
degrees of freedom, chi-square 53
dependent marginals 46
describing groups of numbers 5–23
destroying data 31
differences
  means 98–101
  percentages 101–2
directions
  association 37, 72
  two-way tables 116
disagreement responses 34
dispersion
  means 17
  measures 13–19
  standard deviation 18

distribution
  income 20
  normal curves 96
  sample means 94

'elitism' article 62
epsilon (percentage differences) 29, 47
exam performance tables *see* school performance . . .
expected frequencies 47
  chi-square 52
explanation function, data 107

factors, weighting 139–40
figure-phobias 1
Fisher's exact test 145
flat oval graphs 76
frequency
  expected frequencies 47, 52
  means 13, 94
  tables 6, 12–13
frequency tables
  means 13
  standard deviation 18, 21–2
further reading 147–9
future prediction 72, 82–5, 88, 117
  see also trends

graphs
  correlation logic 73
  flat ovals 76
  presentation 106–29
  quadrants 73–7
  slopes 84–5
  squeezed ovals 75
  symmetry 95
  or tables 107–8
  see also scattergraphs
grouped bar charts 123–4

higher level two-way tables 29–31
histograms 126
  see also bar charts

income, distribution 20
independence, associations 25–31
independent marginals 27
independent variables 25
index numbers 34, 35, 42
indices *see* index numbers
inter-quartile ranges 16–17
intercepts, axes 83–5
intervening variables 61
italics, use 115, 117

layout, tables 112–14
league tables 7
  association 33–43
  ordinal data 33–9
  presentation 111–12
least squares regression 83
levels, unreliability 48
levels of measurement 7–9
limitations
  analysis 146–7
  linear correlations 79–81
  linear regression 90
  quantitative data analysis 147
  statistics 143, 146–7
line bar charts 125–6
line of best fit 83
line graphs 126–8
linear correlation limitations 79–81
linear regression 82–5
  calculation 92n
  limitations 90
lines, as separators 115
literature review, reports 131
log-linear modelling 146
  further reading 149

marginals
  independent 27
  tables 26, 117
means
  checking differences 98–101
  differences 93–105
  directions 102

dispersion 17
distribution 94
estimation limitation 103
frequencies 13, 94
measurement levels 9–12
moving averages 89, 90
reliability estimation 93–105
scatter 17
tables 13
measurement levels 7–9
averages 9–12
measures *see individual measures*
measuring reliability 46–51
category data 46–9
ordinal data 49–51
Spearman's rank correlation 49
medians
cumulative percentages 12
inter-quartile ranges 16–17
league tables 10
midpoints 14
model of no association 46
modes 9, 10
mortality index 40
moving averages 89, 90
multi-stage sampling 139
multiple regression 146, 149
multiplication, negative numbers 74
multivariate analysis 25, 60–9

negative correlation 72, 76, 134
negative number multiplication 74
negative rank correlation 36
negative slopes 85
nominal variables 7–9, 15
non-probability sampling 140–1
non-random sampling 136–40
nonparametric tests 80
normal curves 94, 95–6
'Northern Renaissance' article 11

one-tail tests 78–9, 102
ordinal data
examples 38
league tables 33–9
measuring reliability 49–51
rank correlation 33–9
ordinal variables 7–9, 15
ordinals *see* ordinal variables
outliers 81
oval graphs 75
oval scattergrams 74

parameters (data spread restriction) 80–1
path analysis 66, 68
causal modelling 146
further reading 149
Pearson's r correlation 72, 76–8, 80–1
minimum values 91
outliers effect 81
reliability 91
'pensions' article 86–7
percentage differences 29, 101–2
perfect correlations 36, 76
performance
case studies 40
index numbers 34
tables 40
pie charts 120–1
pivots 83–4
population
definitions 136–7
random samples 44
ranges 93–105
samples 44
positive correlation 76
positive slopes 85
predicting the future 72, 82–5, 88
presentation
data 106–29
graphs 106–29
league tables 111–12
tables 106–29

INDEX

prior variables, relationships 61
probability sampling 136–9
   types 138
product moment correlation 76–7, 79
   see also Pearson's *r* correlation
projections *see* future prediction

quadrants
   graphs 73–7
   signs 75
quantitative data analysis, limitations 147
quasi-market mechanisms 132
questionnaires 9–10

random sampling 136–40
   populations 44
   simple 137–8
ranges 16–17
   population 93–105
rank correlation 33–43
   case studies 40
   differences 43
   index numbers 42
   interpretation 41
   minimum values 55, 92
   negative 36
   none 36
   perfect 36
   ratio-level numbers 38
   reliability 49, 55–6, 92
   universities 39, 42, 50
rankings, tables 113
ratio 8, 15
ratio-level correlation 38, 81
ratio-level numbers 38
reference function 107
regression analysis 86–90
relationships
   intervening variables 61
   prior variables 61
   sets of numbers 72–92

reliability
   association confusion 45
   association differences 58
   averages 93–105
   chi-square 54
   measurement 46–51
   Pearson's *r* 91
   rank correlation 55–6
   results 44–56
   $r(rho)_s$ correlation 55–6
   scattergraphs 78
   small samples 144–5
   Spearman's rho 92
reliable nonsense 147
reports
   background information 131
   components 131
   literature review 131
   statistical significance 118
   validity of measures 131
research methods 133
residuals 47, 52–3
responses, agreements 34
results reliability 44–56
'reverse logic', confidence intervals 99–100
rounding off numbers 114
rows
   aggregates 115
   arrangements 114–15
   marginals 117
$r(rho)_s$ correlation 37
   calculation 41–2
   reliability 55–6
   see also rank correlation
rules, category tables 48–9

sampling
   frames 137
   probability sampling 136–9
   techniques 130, 135–41
   weighting factors 139–40
scatter, means 17

scattergrams *see* scattergraphs
scattergraphs 72–9, 126–8, 134
    one-tail tests 78–9
    reliability 78
school performance tables 130–5
separators, horizontal lines 115
significance testing 44–56
simple random sampling 137–8
size reduction
    data 114–18
    tables 118
skewness, measures 19–21
slopes, graphs 84–5
small samples
    reliability 145
    standard deviation 144
Spearman's rho *see* rank correlation
squeezed oval graphs 75
stacked bar charts 124–5
standard deviation
    dispersion 18
    frequency tables 18, 21–2
    small samples 145
standard error 95, 100
standardised scores 104
standardised tables 66–9
    path analysis 66, 68
statistical significance 45
    reporting 118
    see also reliability
statistical tests, summary 143
storage, data 107
stratified sampling 138
strength of association 72
structure, tables 108–11
style, tables 112–14
subtotals *see* marginals
summary numbers 5
symmetrical measures 58
symmetry, graphs 95

t distribution 145
t test *see t* distribution

tables 4
    cells 26–7
    content 112
    contingency 32
    frequency 6
    or graphs 107–8
    guidelines 108–18
    layout 112–14
    league tables 111–12
    performance 40
    presentation 106–29
    projections 117
    rankings 113
    size reduction 118
    standardised 66–9
    structure 108–11
    style 112–14
    two-way 27–31
tests, summary 143
three-way tables 109–10
total sample size 115
trends, regression analysis 86–90
two-way tables 109–10
    analysis 27–9
    categories 30
    directions 116
    higher levels 29–31

universities
    applications 66–9
    index numbers 35
    rank correlation 39, 42, 50
unreliability, levels 48

V measures *see* Cramer V
validity of reports 131
variables 6
    association between 24
    control 61
    dependency 25
    nominal 7–9, 15

ordinal 7–9, 15
ratio 8
variance 18–19
   see also standard deviation

Web address 148
weighting factors 139–40

'women's pay' article 99

Yates's correction 145

Z scores 104
zero correlation 76, 77